David Payne and Sue Jennings

geography 360°

Core Book 3

www.heinemann.co.uk

✓ Free online support
✓ Useful weblinks
✓ 24 hour online ordering

01865 888058

Heinemann Educational Publishers
Halley Court, Jordan Hill, Oxford OX2 8EJ
Part of Harcourt Education

Heinemann is the registered trademark of
Harcourt Education Limited

© Harcourt Education Limited, 2006

First published 2006

11 10 09 08 07 06
10 9 8 7 6 5 4 3 2 1

British Library Cataloguing in Publication Data is available
from the British Library on request.

10-digit ISBN 0 435356 73 9
13-digit ISBN 978 0 435356 73 6

Websites
On pages where you are
asked to go to
www.heinemann.co.uk/hotlinks
to complete a task or download
information, please insert the
code **6739P** at the website.

Edited by Janice Baiton
Designed by hicksdesign and typeset and illustrated by HL Studios
Original illustrations © Harcourt Education Limited, 2006
Printed and bound in Italy by Printer Trento S.r.l.
Cover photo: © Getty Images
Picture research by Ginny Stroud-Lewis

Acknowledgements

Maps and extracts

Page 18 Source A: Lyn Topinka. Page 25 Source E: NI Syndication. Page 48 Source A: This map is a
product of a joint study by the International Water Management Institute (IWMI), the World Resources
Institute (WRI), the Center for Environmental Systems Research of Kassel University (KU) and IUCN –
The World Conservation Union. Page 52 Source A: Food and Agriculture Organisation of the United
Nations. Page 67 Source C: World Bank. Page 82 Sources A and B: Ordnance Survey. Pages 90 and
91: Oxfam GB. Page 103: The Independent. Page 104 Source B: Goddard Institute for Space Studies.
Page 104 Source E: United Nations. Page 111: Oxfam GB. Page 113: FARM-Africa. Page 114 Source
A: World Health Organisation; United Nations. Page 115: WaterAid. Page 116 Source A: World Bank;
United Nations. Page 118 Source A: United Nations. Page 130 Source A: World Vision. Page 144:
Ordnance Survey.

Photos

Pages 33, 37, 69, 70, 73, 74, 78, 79, 105, 113, 124, 127, 134: Alamy Images. Page 60: Art
Directors & Trip. Pages 5, 13, 15, 16, 18, 19, 26, 33, 41, 60, 84, 85, 86, 89, 90, 93, 95, 101,
104, 106, 108, 120, 122, 125, 129, 131, 132, 138: Corbis. Pages 114, 129: David Payne.
Page 47: Digital Vision. Pages 24, 98, 123: Empics. Pages 57, 64: Eye Ubiquitous. Pages 12, 14,
22, 25, 26, 28, 29, 32, 38, 50, 56, 66, 67, 94, 104, 106, 127, 132: Getty Images. Page 54:
ICCE. Page 43: Impact Photos. Page 103: Lonely Planet. Page 76: Mirrorpix. Pages 35, 106: NHPA.
Page 111: Oxfam. Pages 84, 86, 88, 92, 99, 106, 110, 113, 127, 131: Panos Pictures. Page 102:
Photodisc. Pages 73, 74: Rex Features. Pages 11, 16, 20, 24, 59: Science Photo Library. Page 68:
Sergio Dorantes. Pages 20, 46, 49, 51, 52, 54, 58, 59, 61, 63, 83, 97, 108, 121: Still Pictures.
Page 94: Topfoto.co.uk. Page 115: WaterAid.

Every effort has been made to contact copyright holders of material reproduced in this book.
Any omissions will be rectified in subsequent printings if notice is given to the publishers.

Contents

» 1 Living with earthquakes and volcanoes

Earthquakes and volcanic eruptions can have disastrous effects, especially if they happen in places where many people live. Understanding the causes and effects of earthquakes and volcanoes can help to reduce their effects on people and environments.

Learning objectives

What are you going to learn about in this chapter?

> Where earthquakes and volcanoes happen
> Why earthquakes and volcanic eruptions happen
> The effects of earthquakes and volcanoes in different parts of the world
> How the effects of earthquakes and volcanoes can be reduced
> How a tsunami starts
> What it is like to live through a tsunami
> Why aid is needed after a natural disaster
> Why people live in active areas

A Earthquake in Kobe, Japan

What is the earth like?

> Understanding that the earth is not a solid mass
> Finding out that inside the earth it is very hot

The earth is made up of three main layers: the crust, the mantle and the core (**A**). The crust is a thin surface which forms the land on which we live. It floats on the semi-liquid (**molten**) mantle. The core is the centre of the earth and is made of iron.

'Scientists believe that 5,000 million years ago the earth was a molten mass – since then the outer layer has cooled and become solid.'

Plates
Huge blocks of the earth's crust.

The core
Scientists believe that:
• the inner core is solid, because it is so dense
• the outer core is molten rock.

Plate boundary
Where plates meet.

The mantle
The mantle is a layer of molten rock (**magma**) underneath the crust.

The crust
• Where land is on the surface it is called continental crust and is usually 20–60 km thick.
• Where oceans are on the surface it is called oceanic crust and is thinner – usually between 8 and 25 km thick.
• In some places the crust is very thin and molten material bubbles up to the surface through cracks – these are called 'hot-spots'.

A The earth's layers

HOW HOT IS IT INSIDE THE EARTH?

No one really knows the answer to this question, but we do know that inside the earth is hotter than the outside and scientists believe that the temperature at the centre of the earth could be as high as 5,000°C! Source **B** shows some of the evidence which tells us that temperatures are higher inside the earth.

Geothermal Power Station	**Hot Springs, Geysers**	**Volcanic Eruptions**
Uses hot underground rocks to produce hot water and electricity.	Water heated up underground rises to the surface under enormous pressure.	Molten rock from inside the earth comes to the earth's surface.

B Evidence to show that temperatures inside the earth are higher

THE EARTH'S CRUST

The earth's crust is not one solid mass. It is made up of large pieces called plates. **Continental Plates** have land on the surface and **Oceanic Plates** have an ocean on the surface. Currents in the molten rocks underneath the crust slowly move the plates around in different directions. As you can see from map **C**, in some places the plates are pushing together, in other places they are moving apart. Plate boundaries are the places where plates meet and are known as active zones because this is where earthquakes and volcanoes often occur.

C The earth's major plates

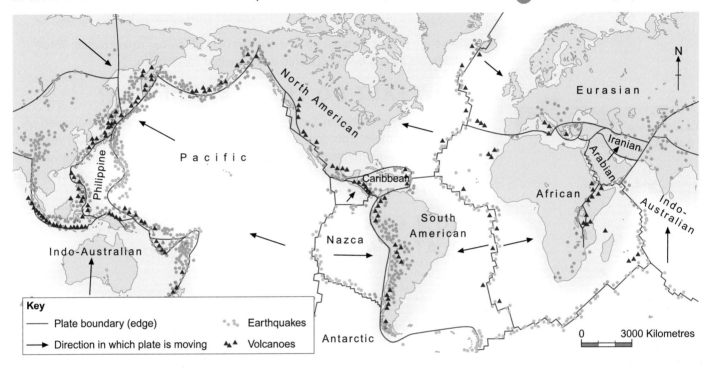

Key
— Plate boundary (edge)　　·.° Earthquakes
→ Direction in which plate is moving　▲▲ Volcanoes

0　　3000 Kilometres

Activities

S **📄**

1　Write a heading: 'What is the earth like?'. Underneath your heading:

  a) Draw a simple diagram to show the structure of the earth. Label the crust, mantle and core.

  b) Write *one* sentence each about the crust, mantle and core.

2　Draw a table like the one below:

How can we tell that it is hot inside the earth?	
Mining	In a deep mine it is much hotter

Add *three* more examples with brief explanations.

3　Look at source **C**.

  a) Which ocean:

    (i) has a ring of earthquakes around the edge?

    (ii) has a line of volcanoes down the middle?

  b) Write a paragraph to describe the location of earthquakes, starting with 'The world map shows that most earthquakes are found ...'. Add some detail and place names (continents/oceans etc.).

Why do earthquakes and volcanoes occur in certain places?

> Understanding that the earth's crust is made up of a number of separate pieces called plates

> Finding out why earthquakes and volcanoes happen near the edges of the earth's plates

The earth's crust is made up of a number of huge pieces called plates. These plates are moved by the currents in the molten rocks below.

HOW FAST DO THE PLATES MOVE?

The plates move between 1 cm and 12 cm a year, which is about the speed of growing fingernails! This does not sound very fast but over thousands of years it can make a big difference to the surface of the earth. About 160 million years ago South America and Africa were next to each other; today they are separated by the Atlantic Ocean (**A**).

HOW CAN WE TELL THAT THE EARTH'S PLATES MOVE?

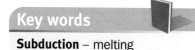

Key words

Subduction – melting

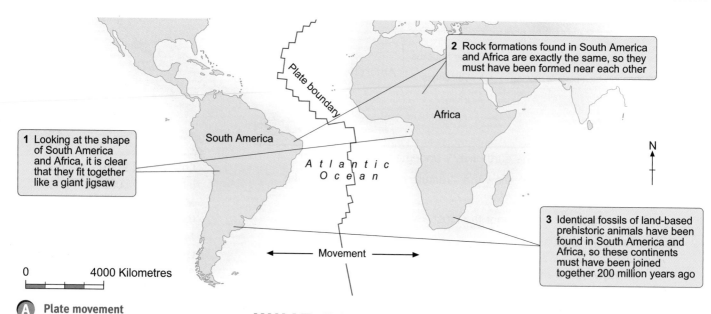

Plate boundary

2 Rock formations found in South America and Africa are exactly the same, so they must have been formed near each other

Africa

1 Looking at the shape of South America and Africa, it is clear that they fit together like a giant jigsaw

South America

A t l a n t i c O c e a n

N

3 Identical fossils of land-based prehistoric animals have been found in South America and Africa, so these continents must have been joined together 200 million years ago

0   4000 Kilometres

← Movement →

A Plate movement

WHAT HAPPENS WHERE PLATES MEET?

The edges of plates are called margins or boundaries (**A**). Plates can move apart, push together or slide past each other. This means that the area where plates meet is very unstable.

WHAT HAPPENS WHEN TWO PLATES MOVE APART?

When two plates move apart molten rock rises into the gap. This quickly cools and forms into ridges of new, solid rock. This can be seen on the bed of the Atlantic Ocean (**A**).

WHAT HAPPENS WHEN PLATES PUSH TOGETHER?

1 In some places where two plates are pushing together one can be forced under the other and pushed down into the mantle. The example in **B** shows the Pacific Plate and the Eurasian Plate pushing together.

B Plates pushing together

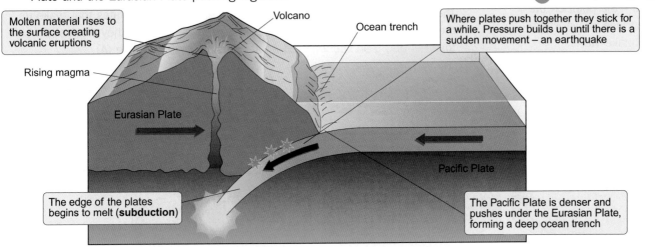

Molten material rises to the surface creating volcanic eruptions

Volcano

Ocean trench

Where plates push together they stick for a while. Pressure builds up until there is a sudden movement – an earthquake

Rising magma

Eurasian Plate

The edge of the plates begins to melt (**subduction**)

Pacific Plate

The Pacific Plate is denser and pushes under the Eurasian Plate, forming a deep ocean trench

2 In other places where two continental plates push together, mountains can be created. The example in **C** shows the formation of the Himalayan mountains, in Asia. **C** Plate movements create mountains

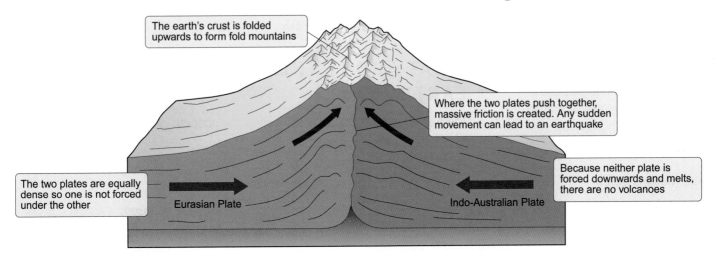

The earth's crust is folded upwards to form fold mountains

Where the two plates push together, massive friction is created. Any sudden movement can lead to an earthquake

Because neither plate is forced downwards and melts, there are no volcanoes

The two plates are equally dense so one is not forced under the other

Eurasian Plate

Indo-Australian Plate

Activities

1 a) Explain how you can tell that South America and Africa were once joined together.

   b) If the plate Britain is on and the North American Plate are moving apart at 6 cm a year, how much further away will Britain be from North America in:

      (i) 10 years?   (ii) 100 years?   (iii) 1,000 years?

2 Draw a sketch to show how the Himalayas are being formed. Don't forget to:

   – include a heading

   – add some notes

   – add some colour.

3 Draw a diagram like the one below.

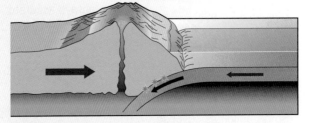

Label your diagram to show what happens when an oceanic plate and a continental plate push together.

What happens in an earthquake?

> Finding out why the earth shakes
> Learning about how earthquakes are measured

The most violent earthquakes occur near the edges of the plates that make up the earth's crust. The plates do not always move smoothly, they often stick together and then suddenly jolt apart, causing an earthquake (**A**).

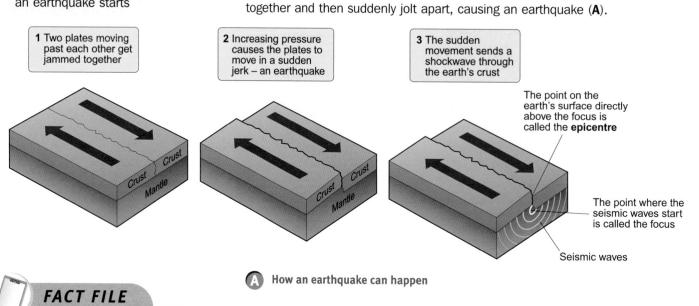

1 Two plates moving past each other get jammed together

2 Increasing pressure causes the plates to move in a sudden jerk – an earthquake

3 The sudden movement sends a shockwave through the earth's crust

The point on the earth's surface directly above the focus is called the **epicentre**

The point where the seismic waves start is called the focus

Seismic waves

A How an earthquake can happen

FACT FILE

The word seismic is Greek for 'shake'. As earthquakes cause the ground to shake, anything to do with earthquakes is called seismic!

WHAT ABOUT THE STRENGTH OF AN EARTHQUAKE?

There are two main scales used to show the strength of an earthquake, both of which are shown below.

The Mercalli scale

Measures the amount of shaking and describes the effects.

1 Not felt

2 Just felt

3 Hanging objects swing

4 Windows rattle – liquids spill

5 Buildings tremble

6 Glass shatters

7 Bricks and tiles fall

8 Walls collapse

9 Some buildings collapse

10 Larger buildings collapse

11 Bridges collapse, railway lines buckle

12 Nearly total destruction

> People have different opinions when describing earthquakes – so the Mercalli scale is not always accurate

The Richter scale

Measures the amount of energy released.

1 Only noticed by instruments

2 Barely felt

3 Slight vibrations

4 Windows rattle, some movement, minor damage

5 Some damage to buildings

6 Walls crack, some buildings collapse

7 Ground cracks – many buildings collapse

8 Large areas destroyed

9 Widespread destruction

> The Richter scale is the most commonly used measurement

> 8.9 Strongest recorded earthquake

> Each additional number is ten times more powerful than the last

HOW ARE EARTHQUAKES MEASURED?

An instrument called a seismometer is used to record the shaking of the earth (photo **B**).

The information collected is shown on a seismograph (**C**). Comparing seismographs from different places can help to tell where an earthquake started.

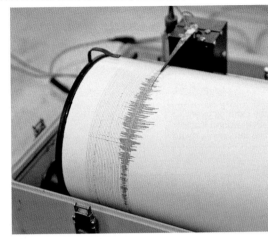

B A seismometer

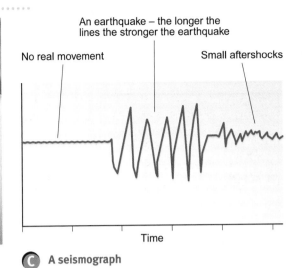

An earthquake – the longer the lines the stronger the earthquake

No real movement

Small aftershocks

Time

C A seismograph

HOW DO EARTHQUAKES CAUSE DAMAGE?

Buildings destroyed by the earthquake

Shaking causes buildings to collapse

Electrical cables damaged causing fires

River banks broken causing flooding

Earth movements cause bridges to collapse

Gas pipes broken causing fires

Large trees fall causing damage

Shaking causes landslides

Activities

1 Why is anything to do with earthquakes called 'seismic'?

2 Look at the diagram on the right which shows part of California, on the west coast of the USA.

a) Explain why California often has earthquakes.

b) What types of damage might a powerful earthquake cause in cities such as San Francisco or Los Angeles?

c) Why might it be difficult to rescue people trapped in buildings after an earthquake?

3 a) Why is the Richter scale most commonly used to describe the power of earthquakes?

b) Use the internet (see Hotlinks page ii) to find out the strength and location of the five most powerful earthquakes ever recorded.

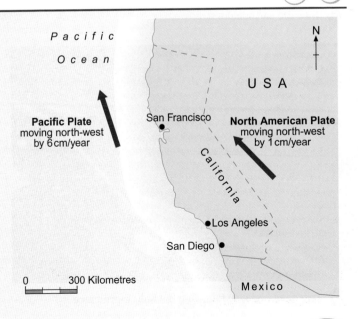

Pacific Ocean

USA

Pacific Plate moving north-west by 6 cm/year

San Francisco

North American Plate moving north-west by 1 cm/year

California

●Los Angeles

San Diego ●

0   300 Kilometres

Mexico

Case study: the Kobe earthquake – Japan

> Understanding the effects of an earthquake
> Finding out how people respond to earthquakes

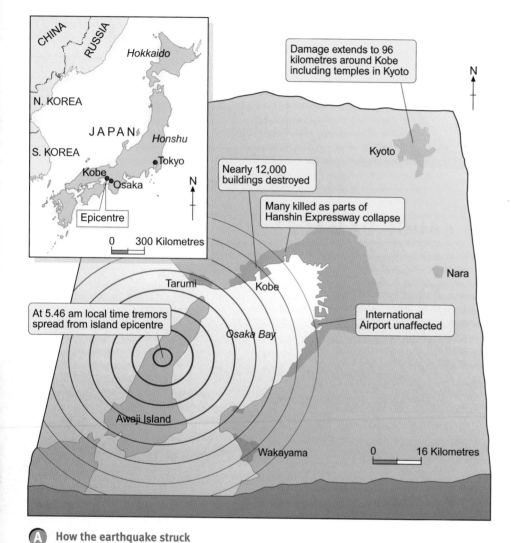

Damage extends to 96 kilometres around Kobe including temples in Kyoto

Nearly 12,000 buildings destroyed

Many killed as parts of Hanshin Expressway collapse

International Airport unaffected

At 5.46 am local time tremors spread from island epicentre

Kyoto

Nara

Tarumi

Kobe

Osaka Bay

Awaji Island

Wakayama

0   16 Kilometres

N

CHINA
RUSSIA
Hokkaido
N. KOREA
JAPAN
Honshu
S. KOREA
Tokyo
Kobe
Osaka
Epicentre
0   300 Kilometres
N

A How the earthquake struck

Kobe is the sixth largest city in Japan and one of the world's largest ports.

At 5.46 am on the 17 January 1995 the city was rocked by a series of earthquake **tremors** (**A**). The worst of these lasted just over 20 seconds and was recorded at 7.2 on the Richter scale.

In a matter of minutes one of the most modern cities in the world had become a disaster area (**B, C**).

FACT FILE

Kobe was Japan's worst earthquake for 72 years

6,310 people were killed

45,000 people were hurt

75,000 buildings were damaged

Rebuilding the city cost over £80 billion

Key words

Tremor – shaking caused by an earthquake

B Damaged highway following the Kobe earthquake

C Damaged buildings following the Kobe earthquake

WHAT WAS IT LIKE TO LIVE THROUGH THE EARTHQUAKE?

The following comments were made by people in the area at the time of the earthquake.

> 'I drive to work on the Hanshin Expressway every morning at about 5.30 am. On the day of the earthquake the car was suddenly thrown across the road and everything seemed to stop. It was only later I realised that parts of the road had collapsed, killing a number of people.'
> **Local factory worker**

> 'It was a quiet morning, like any other. Then suddenly there was a rumbling noise, which got louder and louder and everything started to shake. A lot of buildings collapsed – many people must have been trapped. I will never forget the sight of people desperately moving rubble with their bare hands to try to get to their loved ones.'
> **Local resident**

> 'I was suddenly woken by a tremendous flash, which lit up the sky. I was later told it was an electrical explosion. Everything moved about for what seemed like ages – but was probably only about 20 seconds!'
> **Visiting businessman**

WHAT WAS IT LIKE FOR THE EMERGENCY SERVICES?

The fire, ambulance and police services found it very difficult to cope with the effects of the earthquake for the following reasons:

- Broken gas mains and electrical cables caused hundreds of fires.
- Lots of roads and bridges had been destroyed.
- Many roads were blocked by collapsed buildings.
- Without electricity, communication was more difficult.

REBUILDING THE AREA

The Kobe earthquake destroyed large parts of one of the richest cities in Japan. Many people were surprised that a country like Japan was not better prepared for earthquakes.

Since the earthquake large areas of the city have been rebuilt (**D**) and new measures put in place to make sure that the effects of any future earthquakes are reduced. These include:

- building wider roads with more space between buildings
- using more fire-resistant materials in buildings
- making sure new buildings are earthquake proof
- not allowing building on unstable ground.

D Kobe city rebuilt two years after the earthquake

Activities

1. Write a brief newspaper article about the Kobe earthquake. Use the headline 'The quake that rocked a nation' and include:
   - general background about the area
   - the effect of the earthquake
   - some personal observations from people who were in Kobe at the time.

2. Why were fires a major problem?

3. Why might the effects of the earthquake have been worse if it had happened two hours later?

4. a) Draw an outline sketch of *either* the photograph showing damage to roads (**B**) *or* the one showing damage to buildings (**C**).

   b) Put labels on your sketch to describe the main points. (See page 154 of *SKILLS in geography*.)

5. Explain how any *two* of the new planning measures may make the area safer in the event of a future earthquake.

What happened in the 2003 earthquake in Iran?

> Learning about the effects of an earthquake in a developing country
> Understanding the effects of an earthquake in a developing country

On Friday 26 December 2003 at 5.27 pm an earthquake, measuring 6.3 on the Richter scale, hit the Iranian city of Bam, a city of 80,000 people (**A**). The city is famous for its 2,000-year-old red brick citadel and fortress, which attracts thousands of tourists each year (**B**).

WHAT CAUSED THE EARTHQUAKE?

Iran, on the edge of three major plate boundaries, is a very active earthquake area. On 26 December the Iranian and Arabian plate moved together, causing the massive earthquake (**B**).

The following news reports describe what it was like in Bam after the earthquake.

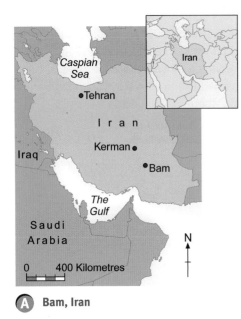

A Bam, Iran

C Bam Citadel before the earthquake

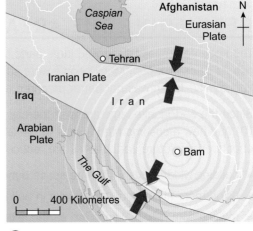

B Plate movement causing the earthquake

D Bam Citadel after the earthquake

QUAKE ROCKS ANCIENT CITY OF BAM

Over 20,000 people were killed yesterday when an earthquake hit the ancient Iranian city of Bam. With thousands of homes destroyed, there are fears that many more will die from being left homeless in the winter cold.

Bam is an ancient city of over 80,000 people, with a number of mud brick buildings over 2,000 years old. It only has two hospitals and both were badly damaged by the earthquake.

City of Bam destroyed in deadly earthquake

Thousands of homes were destroyed when an earthquake hit the ancient Iranian city of Bam yesterday. Over 20,000 people were killed and many more may die because of cold or threat of disease. The two hospitals in the city have been damaged, so people cannot get the help they desperately need.

D Emergency shelter after the earthquake

EARTHQUAKE KILLS THOUSANDS

Just before dawn this morning an earthquake devastated the city of Bam in Iran. Mud brick homes in the city and surrounding villages were reduced to rubble and up to 40,000 people are feared dead. Rescue volunteers, doctors and paramedics are being flown to the country to help survivors many of whom have lost everything.

BAM – ONE YEAR ON

It often takes poor countries a long time to rebuild after an earthquake. The following news report was written a year after the earthquake.

BAM – ONE YEAR LATER

A year after the earthquake in Bam, the city is still trying to recover.

Nearly 70,000 people were left homeless after the quake – over 30,000 of them are still living in temporary shacks. Clearing up the millions of tons of rubble has been an enormous task. The government has used over a thousand trucks non-stop and there are still piles of rubbish everywhere.

A local housing minister said that it costs about £500 to rebuild each house – a lot of money in a poor country. He added that there are still about 20,000 houses, 3,000 shops and many schools and health centres to be rebuilt.

Activities

1   Write a brief report of the earthquake in Bam. Use the title 'The Bam earthquake – Iran 2003'. Mention:
    - where it happened
    - when it happened
    - how powerful it was
    - what the city was like before the earthquake
    - how the earthquake affected people and buildings.

2   Look at the headlines to the newspaper reports. Suggest another two headlines that could be used in reports about the earthquake.

3   Why is it difficult for poor countries to cope after earthquakes?

4   **Extension** Use the internet (see Hotlinks, page ii) to make a table of earthquakes in Iran during the last thirty years. Make sure you give the date, the strength of the earthquake and the number of deaths. What does your table tell you about the pattern of earthquakes in Iran?

What happens when a volcano erupts?

> Learning about different types of erupted material
> Understanding the effects of an erupting volcano

WHAT IS A VOLCANO?

A volcano is an opening or vent in the earth's crust where different materials are able to reach the earth's surface.

WHAT SORTS OF MATERIAL CAN REACH THE EARTH'S SURFACE?

Lots of different types of material can be forced up from inside the earth's crust during a volcanic eruption (**A–D**). Not all volcanic eruptions are explosive. In places like Hawaii **lava** flows in channels and can be studied at quite close range.

Key words

Lava – molten rock on the earth's surface

A Molten lava (liquid rock) can be:
– thick and sticky and move quite slowly
– thin and runny and flow very quickly.

Types of material erupted from the earth

B Steam and volcanic dust often come out of small eruptions.

C Red hot ash is erupted from many volcanoes and is very dangerous.

D Volcanic bombs – large blocks of hot rock – can be thrown hundreds of feet in the air during an eruption.

WHAT DAMAGE CAN A VOLCANIC ERUPTION CAUSE?

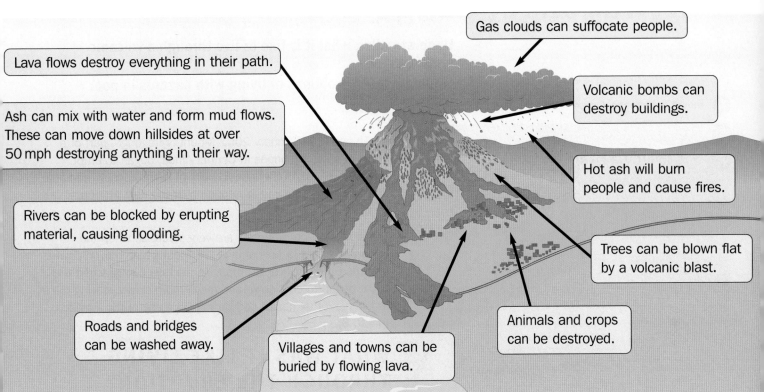

Gas clouds can suffocate people.

Lava flows destroy everything in their path.

Ash can mix with water and form mud flows. These can move down hillsides at over 50 mph destroying anything in their way.

Volcanic bombs can destroy buildings.

Hot ash will burn people and cause fires.

Rivers can be blocked by erupting material, causing flooding.

Trees can be blown flat by a volcanic blast.

Roads and bridges can be washed away.

Villages and towns can be buried by flowing lava.

Animals and crops can be destroyed.

ARE ALL VOLCANOES ACTIVE?

Volcanoes can be active, dormant or extinct.

Active volcanoes   have erupted recently and are expected to erupt again. There are over a thousand active volcanoes, many around the edge of the Pacific Ocean.

Dormant volcanoes   have not erupted for many years but could still erupt. Some of the volcanoes in the western United States are in this category.

Extinct volcanoes   are not expected to erupt again in the future.

Activities

1   What is the difference between magma (see page 7) and lava?

2   Draw a spider diagram to show the different materials that can be erupted from the earth's crust.

3   Write a brief newspaper article describing the possible effects of a volcanic eruption – don't forget to include a headline!

4   a) Why doesn't the UK have any active volcanoes (see page 7)?

    b) Where are the closest active volcanoes to the UK (see page 7)?

5   Use the internet (see Hotlinks, page ii) to locate *five* volcanoes currently erupting.

Case study: the eruption of Mount Nyiragongo, Congo

> Understanding what it is like to live through a volcanic eruption
> Finding out the problems of living with hazards in poor countries

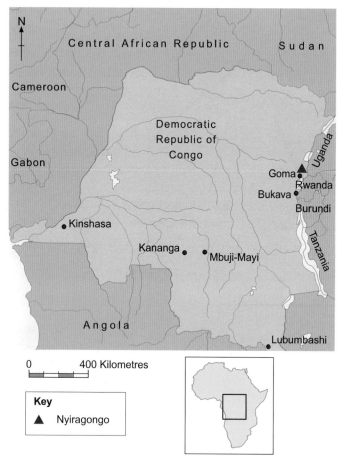

A Mount Nyiragongo, Congo

Source: Developed by Lyn Topinka, Cascades Volcano Observatory

On Thursday 17 January 2002 Mount Nyiragongo, part of a chain of eight volcanoes in Central Africa, began to erupt. Ten miles south, in the lakeside city of Goma, people had little idea of what was going to happen over the next 48 hours (map **A**).

The slopes of the mountain are very steep, so when the volcano erupted, a flow of lava 1.8 metres high moved down towards Goma at over 30 mph. As the lava flow reached the city it began to cool down and move more slowly, leaving thousands of tonnes of cooling rock in many of the streets (photo **B**).

WHAT WAS IT LIKE LIVING THROUGH THE ERUPTION?

The following newspaper articles show what it was like to live through the eruption.

THOUSANDS FLEE FROM RIVER OF LAVA

A volcanic eruption lit up the African city of Goma last night as a river of lava poured from the Nyiragongo volcano. It destroyed villages in its path, turning buildings to ash and forcing thousands of people to flee their homes. By this morning nearly 300,000 people had moved away. The city of Goma was like a ghost town with hundreds of shacks and houses and the catholic cathedral destroyed. In the main shopping area, lava flowed down the streets, giving off gases that stung the eyes and burnt the lungs. On the edge of the city, the airport runway had been destroyed and many buildings in the port were on fire.

 Devastation in the main street of Goma

We thought this would never happen

The volcanic eruption has left thousands in desperate need of food, water and shelter

Aid agencies are preparing for the worst in Goma, where over half the buildings have been destroyed. A foreign charity worker said, 'We have to find people shelter. There is no electricity or running water, the water pumping stations have been destroyed.

Our food warehouses and trucks have been destroyed, main roads and the airport are closed: things look desperate.' A government spokesman said, 'The situation is very bad – we need food, water, medicine, tents and so on. We did not think the volcano would ever cause so much damage, but it has. Whole neighbourhoods have been burnt to the ground.'

 Residents of Goma carrying water rations

PEOPLE RETURN TO THEIR HOMES IN GOMA

Three days after the eruption, people began to return to their homes in Goma. For some, their homes had completely gone, while others had the additional problem of finding that their possessions had been stolen.

Aid teams are trying to repair the water system and are also giving out water purification tablets. Tankers of water and food parcels have been sent to the worst affected areas.

D **Women of Goma cooking meals outside buildings being used for shelter**

Activities

1   a) Why was Goma affected by an eruption that happened ten miles away?

    b) Why were only a small number of people killed in the eruption?

2   Imagine that you are a resident of Goma and are writing to a friend in England. Describe what living through the eruption was like.

3   Explain the following two statements made by aid workers.

    'Getting the airport open is a major priority.'

    'If we don't get clean water to the people the situation will get much worse.'

4   Why will it take Goma a long time to get back to normal?

How can earthquakes and volcanoes be made less of a hazard?

> Understanding that prediction, planning and preparation can reduce risks

> Learning about some of the methods used to reduce the risks of earthquakes and volcanoes

CAN EARTHQUAKES BE PREDICTED?

- Most major earthquakes happen at plate boundaries, so using instruments (photo **A**) to detect small movements in these areas might give a clue to a future earthquake.
- Gas is sometimes released from rocks that have cracked under the ground.
- Underground water levels sometimes change before an earthquake.
- Small movements can be picked up on a seismometer. This might suggest that a larger movement is on the way.
- It is often said that animals behave strangely before an earthquake.

A A laser detector

Emergency planning officer

'It's not easy but there are things we can look out for and do.'

PLANNING FOR EARTHQUAKES

The following measures can reduce the risks from earthquakes.

- Make sure that bridges and elevated roads are strengthened.
- Use building materials that don't burn easily.
- Put services like gas and electricity in flexible pipes that bend but don't break.
- Leave bigger spaces between buildings for emergency vehicles.

'The biggest danger in an earthquake is caused by buildings collapsing or catching fire. It is possible to construct buildings that are earthquake proof – The TransAmerica Pyramid in San Francisco is a good example of this.'

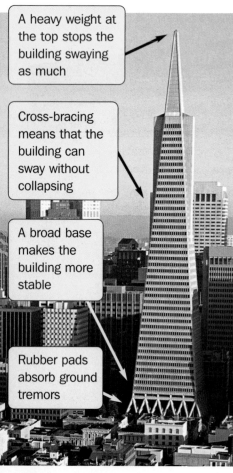

A heavy weight at the top stops the building swaying as much

Cross-bracing means that the building can sway without collapsing

A broad base makes the building more stable

Rubber pads absorb ground tremors

B TransAmerica Pyramid, San Francisco

WHY IS PREPARATION IMPORTANT IN AN EARTHQUAKE ZONE?

'Preparation is not just making sure that the ambulance, hospital, fire and police service know what to do – it is also about preparing individual people. This could save your life.'

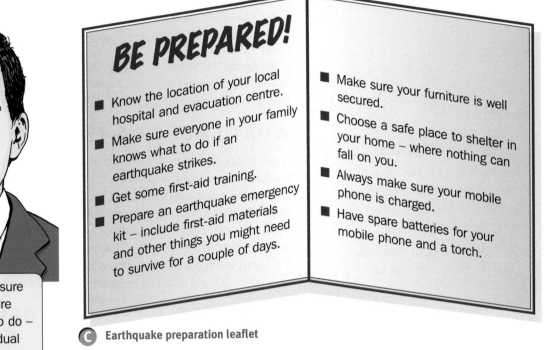

BE PREPARED!

- Know the location of your local hospital and evacuation centre.
- Make sure everyone in your family knows what to do if an earthquake strikes.
- Get some first-aid training.
- Prepare an earthquake emergency kit – include first-aid materials and other things you might need to survive for a couple of days.

- Make sure your furniture is well secured.
- Choose a safe place to shelter in your home – where nothing can fall on you.
- Always make sure your mobile phone is charged.
- Have spare batteries for your mobile phone and a torch.

C Earthquake preparation leaflet

WHAT ABOUT VOLCANOES?

Before volcanoes erupt there are a number of tell-tale signs including:

- small earth tremors
- the side of a volcano begins to bulge or cracks appear
- an increase in temperature, causing ice to melt on volcanic mountain peaks
- small eruptions giving off heat, ash or gas.

If you know an eruption is likely, emergency plans can be put in place. These might include:

- moving people a safe distance away
- preparing shelter for people who have to move
- making sure people have food, water and warm clothes
- setting up emergency transportation and hospitals.

Activities

1 What is meant by:
   - Prediction?
   - Planning?
   - Preparation?

2 Copy out and complete the spider diagram to suggest some of the warning signs before an earthquake.

Warning signs

3 a) What problems might falling buildings create for the emergency services after an earthquake?

   b) How can planning reduce the effects of earthquakes?

4 *Rescue worker*: 'A radio, torch and mobile phone are probably the most useful things to have after an earthquake.'

   Explain the point made by the rescue worker.

5 Why might it be difficult to evacuate people from an area threatened by a volcanic eruption?

What is a tsunami?

> > Understanding what causes a tsunami
> > Finding out what can be done to reduce the effects of a tsunami

In Japanese the word 'tsu' means wave and 'nami' means harbour, so the word 'tsunami' really means 'harbour wave'. It was called this because of giant waves hitting the Japanese coast.

WHAT CAUSES A TSUNAMI?

A tsunami is started by a sudden movement on the seabed, usually caused by an earthquake or volcanic eruption. It is like dropping a heavy object in a bowl of water, causing the water level to rise and create a wave. In a tsunami the whole of the ocean is moving. In deep water tsunamis can move at speeds of over 500 miles per hour and reach places thousands of miles away. So an earthquake in one place can affect another place in a totally different part of the world. Where the tsunami starts the waves created are usually only a few metres high but as they reach land they slow down, but get much higher, causing devastation to coastal areas (source **A**).

HOW DOES A TSUNAMI DEVELOP?

FACT FILE

The highest recorded wave created by a tsunami hit Japan in 1921 and was just under 90 metres high!

Over 80 per cent of tsunamis occur in the Pacific Ocean.

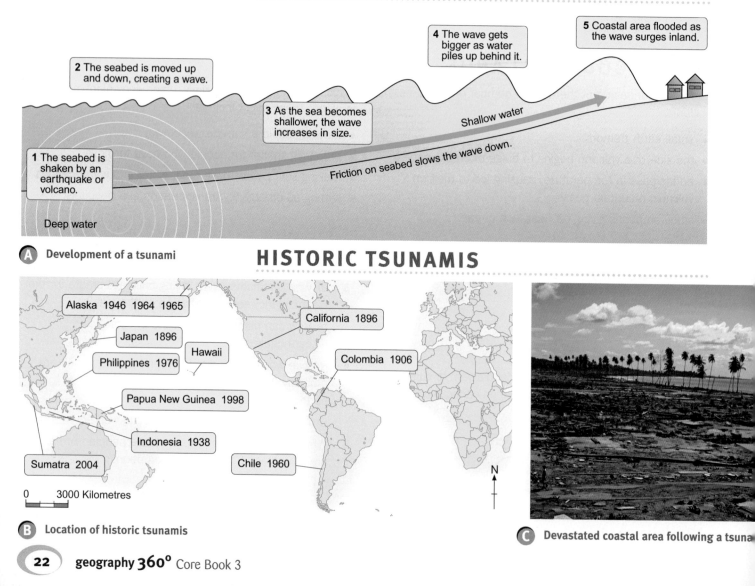

2 The seabed is moved up and down, creating a wave.

4 The wave gets bigger as water piles up behind it.

5 Coastal area flooded as the wave surges inland.

3 As the sea becomes shallower, the wave increases in size.

Shallow water

1 The seabed is shaken by an earthquake or volcano.

Friction on seabed slows the wave down.

Deep water

A Development of a tsunami

HISTORIC TSUNAMIS

Alaska 1946 1964 1965

California 1896

Japan 1896

Hawaii

Philippines 1976

Colombia 1906

Papua New Guinea 1998

Indonesia 1938

Sumatra 2004

Chile 1960

0   3000 Kilometres

N

B Location of historic tsunamis

C Devastated coastal area following a tsuna

HOW CAN YOU PREPARE FOR A TSUNAMI?

In 1948 the Pacific Tsunami Warning System was set up. It is organised from the island of Hawaii (**B**) and involves twenty-four countries in the Pacific Ocean area.

What does the Pacific Tsunami Warning System do?

- Keeps a close eye on earthquake activity
- Records changes in wave patterns and sea levels
- Warns people who might be in danger
- Tells governments when coastal areas need to be evacuated

Not all countries are part of the Pacific warning system, because it is expensive and many of the coutries in the area are poor.

PLANNING FOR A TSUNAMI

In richer parts of the world, such as Japan, coastal areas have been changed to help them cope with tsunamis. The diagrams in source **D** show how a coastal area can be changed so that if a tsunami occurs, the damage will be much less.

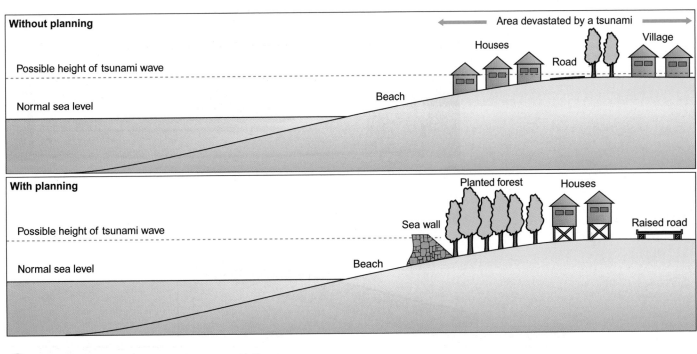

 How planning for a tsunami can help avoid disaster

Activities

1. Draw an annotated diagram to show the development of a tsunami.
2. Describe the scene in photograph **C**, which shows the effects of a tsunami.
3. Why do most tsunamis occur in the Pacific Ocean area? (Look back to pages 7 and 8.)
4. Suggest *three* reasons why a tsunami causes so much damage.
5. Explain how *two* of the measures shown in source **D** might reduce the effect of a tsunami.

Case study: the Indian Ocean tsunami – December 2004

> Understanding that an undersea earthquake can affect places hundreds of miles away
> Finding out about the causes and effects of the Indian Ocean tsunami

On the 26 December 2004 one of the strongest earthquakes ever recorded happened near the coast of north-west Indonesia (photo **A**). The underwater earthquake sent huge waves racing across the Indian Ocean and even reached the coast of East Africa – 4,000 miles away. The earthquake, measuring 8.9 on the Richter scale, was the fifth strongest ever recorded (photo **B**).

A Satellite image showing the developing Indian Ocean tsunami

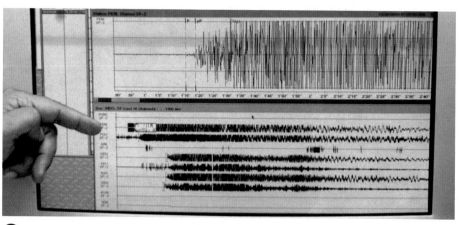

B Seismograph of the earthquake which caused the tsunami

WHAT CAUSED THE TSUNAMI?

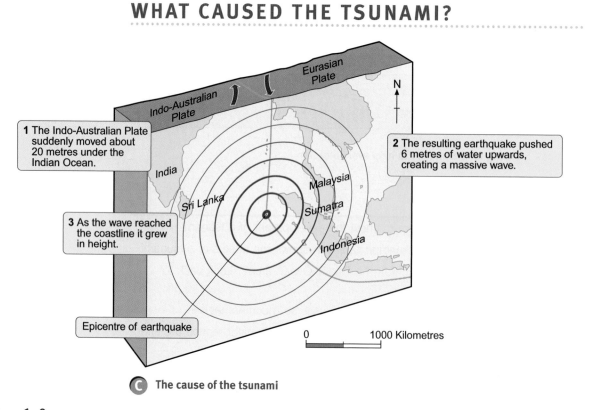

1 The Indo-Australian Plate suddenly moved about 20 metres under the Indian Ocean.

2 The resulting earthquake pushed 6 metres of water upwards, creating a massive wave.

3 As the wave reached the coastline it grew in height.

Epicentre of earthquake

0   1000 Kilometres

C The cause of the tsunami

WHAT WAS THE RESULT IN COASTAL AREAS?

D Tourists fleeing as the tsunami wave approaches

Map **E** shows the immediate effects of the tsunami as massive waves crashed onto the coastal areas surrounding the Indian Ocean.

E Immediate effects of the tsunami

Source: NI Syndication

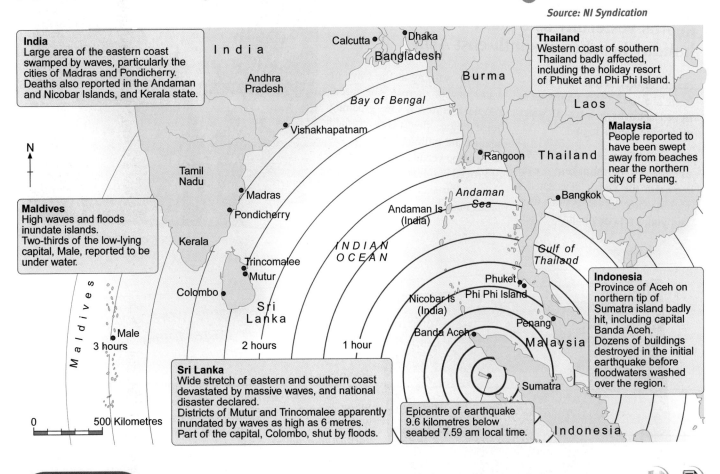

India
Large area of the eastern coast swamped by waves, particularly the cities of Madras and Pondicherry. Deaths also reported in the Andaman and Nicobar Islands, and Kerala state.

Thailand
Western coast of southern Thailand badly affected, including the holiday resort of Phuket and Phi Phi Island.

Malaysia
People reported to have been swept away from beaches near the northern city of Penang.

Maldives
High waves and floods inundate islands. Two-thirds of the low-lying capital, Male, reported to be under water.

Indonesia
Province of Aceh on northern tip of Sumatra island badly hit, including capital Banda Aceh. Dozens of buildings destroyed in the initial earthquake before floodwaters washed over the region.

Sri Lanka
Wide stretch of eastern and southern coast devastated by massive waves, and national disaster declared. Districts of Mutur and Trincomalee apparently inundated by waves as high as 6 metres. Part of the capital, Colombo, shut by floods.

Epicentre of earthquake 9.6 kilometres below seabed 7.59 am local time.

0   500 Kilometres

Activities

1. What caused the tsunami?

2. Why did the tsunami take many people by surprise?

3. How long did it take the waves to reach:
   - the Andaman Islands?
   - Sri Lanka?
   - the Maldives?

4. a) How far is it from the epicentre of the earthquake to Sri Lanka?

   b) Approximately how fast were the waves travelling?

5. Briefly describe the effects of the tsunami in *three* different countries.

Living through the Indian Ocean tsunami

> Learning about how it feels to experience a tsunami
> Understanding that hazards can have both short- and long-term effects

When the tsunami struck the coastal areas surrounding the Indian Ocean, millions of people were affected, including thousands of holidaymakers who had gone to the area for a 'sunshine break' during the Christmas holidays. The following resources give an impression of what it was like to live through the tsunami.

Thousands of people were swept to their deaths yesterday as a giant wave hit the holiday beaches of south-east Asia

Beach resorts across the area – from Thailand to Sri Lanka were ripped apart by a wave of water up to nine metres high.

Generated from one of the largest undersea earthquakes ever reported, the waves devastated a region where 15,000 Britons were enjoying a relaxing holiday. There was little warning as the wave of water swept across the area. People reported a low groaning noise before the waves crashed against the buildings. The waves were so powerful that they totally destroyed buildings and picked up cars and trucks – moving them miles inland.

As millions of people were going about their daily lives, they were totally unaware of the horrors facing them

People were not to know that the gentle swaying of skyscrapers in Singapore was the result of an earthquake which was about to bring a wall of water crashing down on them.

The Indian Ocean does not usually suffer from tsunamis and the area has no warning system. Towns and villages have grown up near the beaches, many based on traditional fishing industries. More recently, the tourist industry has developed with large resorts and holiday homes dotted along the coast. When the wave struck, many of these areas were totally destroyed.

N

0 _____ 1000 Kilometres

INDIA

'We were sitting in our bedroom and heard an enormous roar. Seconds later the door burst open and the room filled up with water. We were swept out of the windows, but managed to struggle towards higher land – and safety.'

THAILAND

'We were on a diving boat and were thrown around by the waves. The boat was taken inland by the wave and stuck between two buildings. We climbed to the roof of a hotel. We were lucky to survive.'

'I was just going for a swim in the pool when a giant wave appeared. The wave knocked me down several times – it was very strong. I was hit by trees, tables and other things, but managed to struggle to safety. I am covered in cuts and bruises, but am lucky to be alive – so many people in the area have lost their lives.'

I n d i a

B u r m a

Thailand

I n d i a n

O c e a n

Sri
Lanka

M a l a y s i a

I n d o n e s i a

SRI LANKA

'There was no warning. The first wave crashed through the buildings – many of which collapsed. Everything was swept along by the waves – even cars and lorries.'

'The water levels are now going down, leaving a scene of total devastation. Most local people have lost their homes.'

(A) Interviews with British holidaymakers affected by the tsunami

Activities (S)

1 Describe the scenes in the photographs on the opposite page.

2 Write a short newspaper report about the tsunami. Include some information about:

- the strength of the tsunami

- some facts and figures

- how it affected people in the area.

3 Why might a lot of people living in the area be affected by the tsunami in the following months?

DISEASE THREATENS A SECOND DISASTER

Aid agencies warned of a second disaster last night – from disease caused by polluted water and food. An aid agency spokeswoman said: 'There are a lot of people drinking contaminated water and eating food picked up from the streets. The chances of disease spreading are very high. Also lots of people have no shelter. There are few doctors and limited drugs; many hospitals have been destroyed.'

Helping people after a natural disaster

> Understanding the importance of aid after a natural disaster
> Finding out about the different types of aid needed after a natural disaster

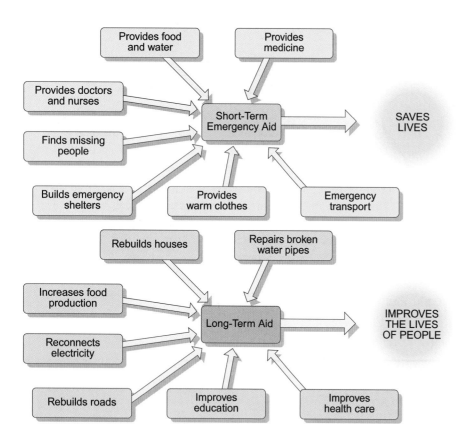

A How aid can help

After a natural disaster countries often face real problems, both immediately and in the longer term. Immediate problems might include shortages of food, water or medical facilities. In the longer term a country could need help to repair roads or rebuild towns and generally get the economy working again.

One way of helping countries is by giving aid (**A**). There are two main types of aid:

1 Government aid: this is money given by one government to another government (**B**).

2 Voluntary aid: this is money given by charities such as Oxfam, the Red Cross/Red Crescent or Christian Aid. Charities are called Non-Governmental Organisations or NGOs.

HOW DID AID HELP PEOPLE AFFECTED BY THE INDIAN OCEAN TSUNAMI?

Government aid

Japan	£260m
USA	£180m
UK	£50m
Sweden	£40m
Spain	£35m
China	£31m
France	£30m
Taiwan	£26m
Australia	£24m
Canada	£17m

B Money given by different countries after the tsunami

1500 American soldiers and 20 helicopters were sent to the area to help distribute food and water.

The Japanese government gave millions of pounds – much of which was used to supply food, shelter and medical help.

Soldiers and heavy machinery from many countries were flown in to help clear roads and airports.

Doctors, nurses and medical equipment were sent to the area from a number of countries.

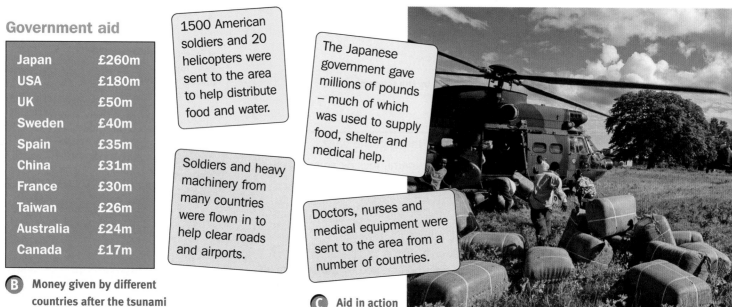

C Aid in action

VOLUNTARY AID

The following diagram shows some of the help given by charities after the Indian Ocean tsunami.

E Voluntary aid in action

INDIA

UNICEF – supplying water purification tablets; medical supplies; blankets

Action Aid – providing clean water; food

Christian Aid – supplying 50,000 emergency packs

THAILAND

World Vision – supplying 2,000 survival kits; arranging transport to take injured people to hospital

SRI LANKA

World Vision – building temporary shelters for 2,000 people

Oxfam – supplying blankets, clothes, food and medical supplies

Red Cross – supplying cooking stoves and water purification tablets; setting up a system to reunite families who have been separated

INDONESIA

World Vision – building shelters for 8,000 people

Oxfam – sending water tanks and materials to build emergency toilets

D Some of the help given by charities after the tsunami

Activities

1. What is meant by:
   - government aid?
   - voluntary aid?

2. Explain the importance of:
   - short-term (emergency) aid
   - long-term aid.

3. Draw a bar graph to show the money given by the top *five* countries after the tsunami – don't forget to label your graph. (See page 148 of *SKILLS in geography*.)

4. Imagine you are a soldier who has been sent to the Indian Ocean area to help after the tsunami. Write a short letter home, describing some of the ways that you are helping.

5. List all the different types of aid given by charities shown in source **D**.

6. Use Google to look up *two* of the charities mentioned on the internet. For each:
   - write down the full name/web address
   - describe *one* short-term/emergency project they are involved with
   - describe *two* long-term projects they are involved with.

Why do people live in active areas?

> > Learning that there are advantages to living in active tectonic areas

> > Understanding that people who live in active areas learn to adapt to them

Looking at a world map it is easy to see that several of the world's largest cities, including Tokyo, Mexico City, Los Angeles and San Francisco are in active earthquake areas. Many people live in areas where there are volcanoes and earthquakes. One reason for this is that serious volcanic eruptions and earthquakes don't happen very often so they feel that they will be safe. After all, if you have lived in an earthquake zone for forty years and have never been affected by a serious earthquake, you probably don't think about it much.

WHY LIVE IN ACTIVE AREAS?

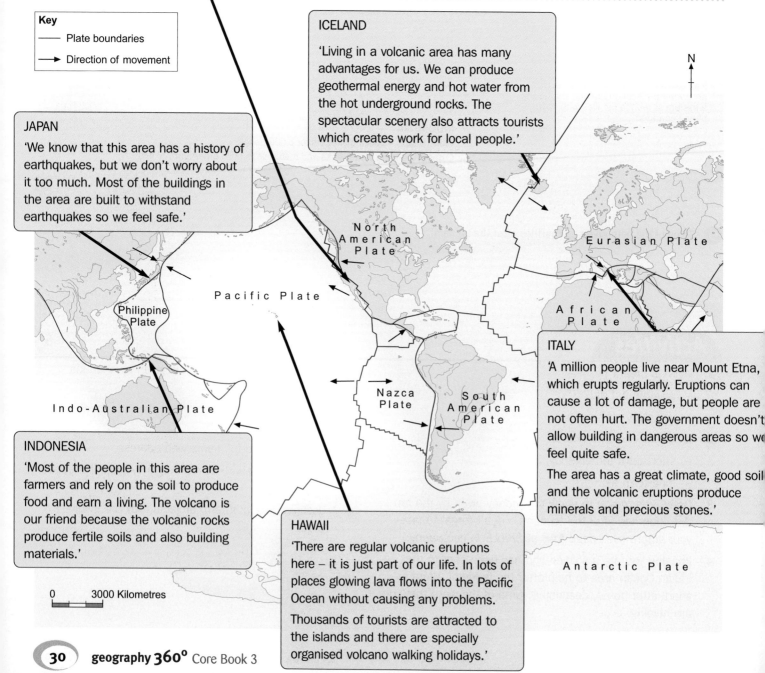

CALIFORNIA (USA)

'There are many advantages to living in this area. The climate here is fantastic, the coast is great for leisure activities and there are plenty of well-paid jobs. We have excellent rescue services if there is a problem.'

Key

—— Plate boundaries

→ Direction of movement

JAPAN

'We know that this area has a history of earthquakes, but we don't worry about it too much. Most of the buildings in the area are built to withstand earthquakes so we feel safe.'

ICELAND

'Living in a volcanic area has many advantages for us. We can produce geothermal energy and hot water from the hot underground rocks. The spectacular scenery also attracts tourists which creates work for local people.'

N

North American Plate

Eurasian Plate

Pacific Plate

Philippine Plate

African Plate

Indo-Australian Plate

Nazca Plate

South American Plate

ITALY

'A million people live near Mount Etna, which erupts regularly. Eruptions can cause a lot of damage, but people are not often hurt. The government doesn't allow building in dangerous areas so we feel quite safe.

The area has a great climate, good soil and the volcanic eruptions produce minerals and precious stones.'

INDONESIA

'Most of the people in this area are farmers and rely on the soil to produce food and earn a living. The volcano is our friend because the volcanic rocks produce fertile soils and also building materials.'

HAWAII

'There are regular volcanic eruptions here – it is just part of our life. In lots of places glowing lava flows into the Pacific Ocean without causing any problems.

Thousands of tourists are attracted to the islands and there are specially organised volcano walking holidays.'

Antarctic Plate

0   3000 Kilometres

The following travel report is from a holiday magazine.

JOURNEY AMONG THE VOLCANOES

Nicaragua is the largest country in Central America, but is still no bigger than England. It is on the edge of two tectonic plates where the earth forces up volcanoes and creates a spectacular scenery of deep valleys and giant waterfalls. The fertile volcanic soils and heavy rainfall mean there is a great variety of plant and animal life. Near the coast of Leon sits Cerro Negro, Nicaragua's newest and most active volcano. Climbing the cone is hot work, but gives fabulous views over the surrounding countryside. Inside the crater the rock is stained with sulphur and other minerals.

Further south a road leads to the edge of the crater of Volcan Masaya, where a sign helpfully tells you to 'shelter under your car in the event of an eruption'! Mombacho volcano, on the outskirts of Granada, is covered in forest with spectacular flowering plants and is a sight not to be missed. The volcanic cone at Cerro Negro was formed in 1998; my guide said you can bike, ski or surf down the hot ash!

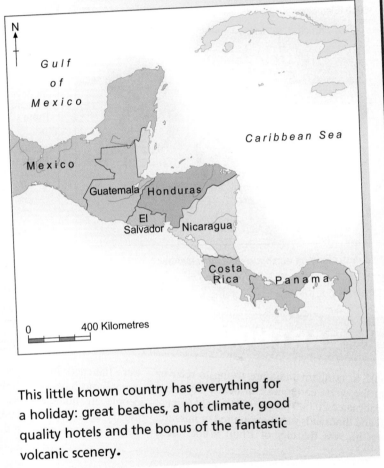

This little known country has everything for a holiday: great beaches, a hot climate, good quality hotels and the bonus of the fantastic volcanic scenery.

Activities

1   Below are two different ways people view living with natural hazards.

    Acceptance: 'Hazards are a part of everyday life. We know they will happen, but live here because the area has so many advantages.'

    Adaptation: 'Hazards can be predicted and warnings given. Modern technology can make areas safer. We can plan so fewer people will be affected.'

    Using examples, explain the two different views about living with hazards.

2   Construct a table to show the advantages of living in active tectonic areas.

3   Imagine you are on holiday in Nicaragua. Write a postcard home describing what it is like and the sort of activities you might be doing.

4   **Research task** Use an atlas, travel brochures or the internet (see Hotlinks, page ii) to identify different tourist locations in volcanic areas.

Living with earthquakes and volcanoes

Key

- Earthquakes
- Volcanoes

1 The location of earthquakes and volcanoes.

  a) Compare the location of earthquakes and volcanoes (**A**) with the position of the earth's plates (**B**).

  b) Why do earthquakes and volcanoes occur at plate boundaries?

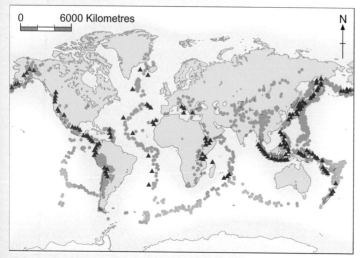

0 — 6000 Kilometres — N

A World map showing earthquakes and volcanoes

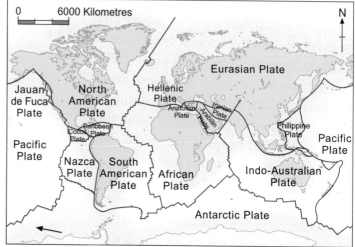

0 — 6000 Kilometres — N

Jauan de Fuca Plate
North American Plate
Hellenic Plate
Anatolian Plate
Iranian Plate
Arabian Plate
Eurasian Plate
Caribbean Plate
Cocos Plate
Pacific Plate
Philippine Plate
Pacific Plate
Nazca Plate
South American Plate
African Plate
Indo-Australian Plate
Antarctic Plate

B World map showing plates

2 The effects of earthquakes and volcanic eruptions.

EARTHQUAKE STRIKES INDIA

People in northern India are trying to recover from the worst earthquake in nearly 50 years. An estimated 20,000 people are thought to have died and thousands more are injured or missing. Worst hit was the city of Bhuj. Whole areas were flattened. In villages surrounding the city many farms have been destroyed. Thousands of people are homeless and emergency services are trying to provide food, water and medical help. There is a growing threat of disease.

C Indian earthquake – newspaper report

Using source **C**:

a) Describe the immediate effects of the Indian earthquake.

b) Suggest what the longer-term effects of the earthquake might be.

c) Why are poor countries often more badly affected by earthquakes?

3 Reducing the effects of earthquakes and volcanoes.

  a) How could the effects of earthquakes be reduced by:

  - Prediction?
  - Planning?
  - Preparation?

  b) List *four* things you might include in an emergency earthquake kit. Explain your choices.

4 The importance of aid after natural disasters.

  a) What sorts of emergency aid are often helpful immediately after an earthquake or volcanic eruption?

  b) What types of long-term aid might be helpful to a developing country recovering from the effects of an earthquake or volcanic eruption?

DEC Emergency Appeal

Tsunami Earthquake

You can help
Donate here

or go online: www.dec.org.uk
or call: 0870 60 60 900

Thank you!

Disasters Emergency Committee

» 2 Ecosystems

Living and non-living things interact and adapt together in these natural ecosystems. They look permanent but they are both easily damaged and easily changed.

Learning objectives

What are you going to learn about in this chapter?

> The workings of an ecosystem

> Different ecosystems and why they are fragile

> How and why people change ecosystems

A Temperate rainforest, USA

B Antarctica

What is an ecosystem?

> Finding what an ecosystem is
> Understanding how an ecosystem works

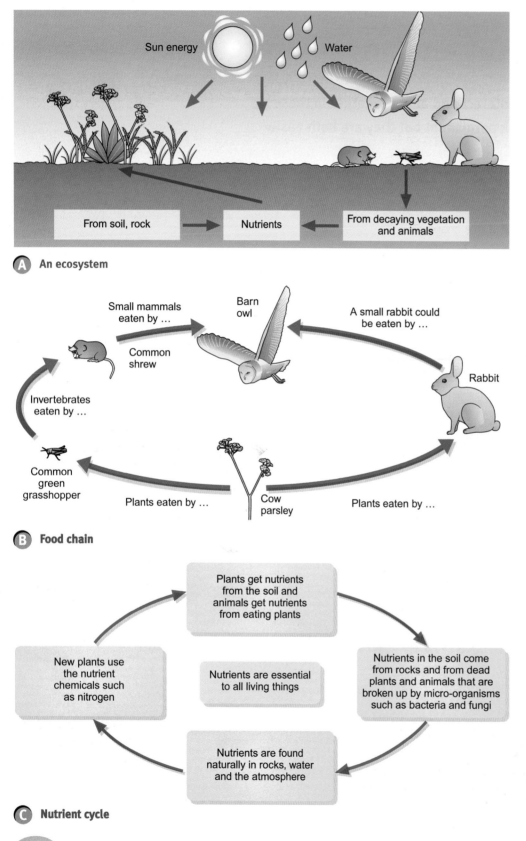

Sun energy

Water

From soil, rock → Nutrients ← From decaying vegetation and animals

A An ecosystem

Small mammals eaten by ...

Barn owl

A small rabbit could be eaten by ...

Common shrew

Rabbit

Invertebrates eaten by ...

Common green grasshopper

Plants eaten by ...

Cow parsley

Plants eaten by ...

B Food chain

Plants get nutrients from the soil and animals get nutrients from eating plants

New plants use the nutrient chemicals such as nitrogen

Nutrients are essential to all living things

Nutrients in the soil come from rocks and from dead plants and animals that are broken up by micro-organisms such as bacteria and fungi

Nutrients are found naturally in rocks, water and the atmosphere

C Nutrient cycle

An **ecosystem** (**A**) is a natural system where everything works together. Living things (plants, animals, insects and micro-organisms) interact with non-living parts of the system (light, heat, water, air, soil and rocks). Ecosystems can be any size – as big as a desert such as the Sahara in Africa, or as small as a pond or a hedgerow. All the parts of an ecosystem depend on each other; this is known as **interdependence**. A change in one part will affect the whole system.

THE FOOD CHAIN IN AN ECOSYSTEM

- Energy in the system comes from the sun and green plants use the sun's energy to **photosynthesise** and make food.
- Animals and insects called herbivores eat the plants and they are then eaten by meat-eating animals, called carnivores.
- Energy moves through the system along this food chain (**B**).

A NUTRIENT CYCLE IN AN ECOSYSTEM

Nutrient chemicals such as nitrogen in the soil are taken up by plants through their roots. When leaves fall or the plants die, the chemicals are returned to the soil, completing the **nutrient cycle** (**C**).

WHY DO SOME ECOSYSTEMS HAVE MORE PLANTS AND ANIMALS THAN OTHERS?

Inputs are things that go into an ecosystem such as heat, light, water, energy and minerals. An ecosystem with a lot of inputs will have a great number of plants and animals living in it. Tropical rainforests, rich in heat, water and nutrients, are home to half the earth's plant and animal species but deserts, with limited water and soil nutrients, have a small range of vegetation and wildlife.

CAN ECOSYSTEMS CHANGE?

Changes in ecosystems can occur naturally or as a result of things that people do. A **drought** causes a natural change in an ecosystem and plants and animals may die as a result. Man-made changes to ecosystems can be large scale, such as when forests are cut down, or they may be small scale, such as cutting a hedge too low before birds have nested.

WHAT IS A BIOME?

D Cactuses and camels have adapted to drought in their ecosystems

A very large ecosystem, such as a hot desert or a tropical rainforest, is called a **biome**. Each biome will have similar living things (plants and animals) and non-living things (soils, rain, sunlight and heat) so a tropical rainforest in South America will have similar climate, plants and animals to one in Africa. The ecosystem or biome in southern England and Europe hundreds of years ago was **deciduous** woodland but there is little woodland left as most of it has been cut down for timber or to clear the land for farming and people.

Activities

1. a) What is an ecosystem? To explain this, either draw a well-labelled diagram (you could make up your own flow diagram) or write a description. For more information you could search the internet (see Hotlinks, page ii), using 'ecosystems' and 'biomes' as your search words.

2. a) Put the following parts of a hedgerow food chain in the right order:

Shrew (eats insects)	Barn owl (eats meat)
Beetle (eats leaves)	Vegetation

   b) Why is the food chain called a 'chain'?

   c) What do you think might happen if barn owls stopped nesting near hedges?

3. a) Why is the nutrient cycle in an ecosystem called a cycle?

   b) Where do shrews, beetles, barn owls and vegetation get their nutrients in the hedgerow ecosystem?

4. Why does a hot desert ecosystem have fewer plants and animals in it than a tropical rainforest? (You could draw two diagrams of the two ecosystems with inputs and outputs labelled.)

5. Find out how camels have adapted to drought.

The tropical rainforest ecosystem

> Finding out about the tropical rainforest ecosystem
> Understanding why tropical rainforests are so important

0 6000 Kilometres

N

Amazon rainforest

Rainforests of Central America

Congo River Basin rainforest

Madagascar rainforest

Rainforests of south-east Asia

WHERE ARE TROPICAL RAINFORESTS?

Tropical rainforests grow in hot humid areas around the equator in South America, Africa and south-east Asia where rainfall is high all year. They cover only a small part of the earth's surface but are very complex ecosystems with more plant and animal species per hectare than anywhere else in the world.

WHAT IS A RAINFOREST LIKE?

Emergent layer
Giant trees like mahogany are spaced wide apart and have mushroom-shaped crowns to get most light (but they are also exposed to strong winds). Although they are very tall, their root systems are very shallow in order to collect nutrients from the top few centimetres of soil so they have **buttress roots** to support them which spread out 10 metres.

Canopy layer
A dense canopy of trees about 30 metres high receives most of the remaining light. Trees are tied together with vines (called lianas) and other plants (called epiphytes) live on them. The canopy is home to 90 per cent of the organisms found in the rainforest, including monkeys, snakes and birds.

Under canopy
Only about 2–15 per cent of the light gets to the under canopy, which has comparatively few plants but lots of straight tree trunks.

Shrub layer
The shrub layer has a dense growth of shrubs and ferns that need little light as well as saplings of the canopy and emergent plants growing to reach the light.

Forest floor
With less than 2 per cent of the light, little grows on the forest floor.

Emergent layer (40–80 metres)

Canopy layer (20–40 metres)

Under canopy (10–20 metres)

Shrub layer (under 10 metres)

Forest floor

WHY ARE TROPICAL RAINFORESTS SO SPECIAL?

- Trees make up 70 per cent of the plants and they produce a lot of the earth's oxygen.

- Plants adjust to the very heavy rainfall with adaptations such as leaves with drip tips to make the water run off quickly. Fungus might grow on the leaves and attack the plant if the water did not run off quickly, as forests are so hot and humid.

- Equatorial climate areas have no clear seasons, as opposed to other areas with clearly defined seasons, so one plant may be in flower and another may be shedding leaves.

- The heavy rain washes nutrients from the soil so most of the nutrients are in the vegetation and roots (about 90 per cent). The nutrients from plants and animals which fall to the forest floor are instantly recycled, so if the trees are cleared there is an immediate loss of nutrients. Other plants do not grow well on the poor soils of cleared rainforest land and the rainforest will not regrow.

B Amazon rainforest

WHY ARE TROPICAL RAINFORESTS SO IMPORTANT?

- Tropical rainforests help to maintain global rain and weather patterns. Most of the rain that falls on the forest evaporates and falls again as rain. If the rainforest is cleared, the rain runs into rivers and away.

- Tropical rainforest plants use carbon dioxide and give out oxygen.

- Over half the earth's plant and animal species are in the rainforests.

- About a quarter of all medicines we use come from rainforest plants. For example, quinine from the cinchona tree is used to treat malaria.

- We use lots of rainforest products in our daily lives including hardwood timbers, rubber and bamboo. Chicle (the original base for chewing gum), cacao, sugar cane and vanilla also come from these areas.

Key words

Buttress roots – roots stretching above ground from a tall tree to help support it

Activities

1. Draw a diagram to show rainforest layers. Use the following as a key and write the numbers on the correct part of your diagram:

   1 emergent layer
   2 canopy, dense tree tops
   3 most animals in trees
   4 under canopy, deep shade
   5 forest floor, 2 per cent of daylight
   6 buttress roots
   7 thin soil layer

2. a) Use the following details and in about 100 words describe the climate in the Amazon rainforest:

   lowest month temperature 27°C
   highest 29°C   all others 28°C
   most rainfall in a month 270 mm
   least 40 mm   total 1900 mm

   b) Choose *five* reasons to explain why rainforests are so important to people.

   c) Use the internet (see Hotlinks, page ii) to research one rainforest product (use 'tropical rainforest' when searching not just 'rainforest'). Find out as much information as you can about where it comes from, who uses it, how they use it and what they use it for. You could present the information on an A3 sheet for display or as an ICT presentation.

Deforestation

> Finding out why rainforests are being cut down
> Thinking about who is affected

WHY IS THE AMAZON FOREST BEING CUT DOWN?

For beef – there were 57 million cattle on ex-forest land in 2002. Five square metres of forest land are needed for every $\frac{1}{4}$ lb burger.

Soya bean is grown on ex-forest land for export to Europe, China, Japan and the USA.

Hardwood trees are cut down for timber; many companies are foreign. Profits go out of the country.

A new road built in 2004 gives people easy access to clear miles each side of the road.

For mines (iron, bauxite, gold) and new industries.

Poor farmers from other parts of Brazil clear the forest to grow crops.

Indian tribes clear the forest to grow food; ash from burnt wood makes the soil fertile, for a time – this is called **slash and burn**.

A Deforestation of the rainforest

FACT FILE

One hectare of rainforest is cut down every second

An area bigger than New York is cut down each day

137 plant, animal and insect species disappear every day, 50,000 a year

25 per cent of Amazon rainforest has been destroyed since 1970

Only 55 per cent of world rainforest remains

2004 saw the most deforestation for ten years

WHAT IS LOST WITH DEFORESTATION?

- Plant, animal and insect species are lost forever and the forest will never naturally regrow. Some trees are hundreds of years old.
- There is loss of way of life for indigenous people.
- Rainforests should contribute significant amounts of global oxygen, so where will that come from?
- CO_2, a major contributor to global warming, is absorbed by rainforests so if less is absorbed, global warming will increase.

THE PHILIPPINES RAINFOREST

In December 2004 decades of **illegal logging** were blamed for the high death toll caused by severe storms.

Manila – 'Decades of illegal logging have made flash floods and landslides deadlier', an official said following a fierce storm that killed 340 people. 'Many victims died after being washed away by swiftly cascading soil and logs when most residents were asleep.'

One spokesperson said, 'Yes, the real culprit is the mud and landslide. Rain caused the flood but there was no forest cover left to hold the mud. Mudslides have washed away roads, bridges and submerged whole towns. We will put a stop to rampant deforestation.'

Philippines

'This disaster is a direct result of the denudation of the forests', another spokesman said.

In the middle of February it was reported, 'Illegal logging activities are continuing. At least four incidents of mahogany timber poaching have been monitored since 1 February.'

The president said, 'We should go after illegal loggers and other poachers of the environment as public enemies with kidnappers and drug lords.'

'Only 13 per cent of the rainforest is left', said an environmentalist, 'it could be gone in fifteen years.'

WHAT CAN BE DONE TO HELP RAINFORESTS SURVIVE?

In March 2005 industrialised countries agreed to control the illegal logging of forests by only buying timber from legal sources.

Harvesting nuts, fruits, oil-producing and medicinal plants from the forest could make more money per acre than growing soya or keeping cattle on deforested land.

The Forest Stewardship Council sets standards for global forest management. Wood can be harvested sustainably and replanted (afforestation).

Activities

1 Design a leaflet or write a letter to a newspaper to tell people why deforestation of the tropical rainforest should stop. Think about who gains from deforestation and who loses, as well as who gets the money from logging, the environmental problems created and what plants and animals are lost with deforestation. For more information you could search the internet (see Hotlinks, page ii) using 'deforestation Brazil' as your search term.

2 How have decades of illegal logging in the Philippines rainforest made flash floods deadlier?

3 Why are illegal loggers regarded as public enemies?

4 What is timber poaching?

Antarctica – a fragile environment

Antarctica

The frozen continent at the earth's South Pole is the driest, coldest, windiest place in the world. The coldest temperature recorded was −89.4°C. The climate is so cold nothing rots or decomposes. Seals killed by the first explorers can still be seen in some places. The strongest winds are up to 200 mph. This ice desert has less rainfall than the Sahara but 70 per cent of the freshwater in the world is frozen in the ice sheet. Average temperatures are −34°C in winter and −7°C in summer. The winter is dark, 24 hours of darkness when the South Pole is tilted away from the sun in June, but the days are long in summer with up to 24 hours of daylight in December.

ANTARCTICA'S INHABITANTS

Only a few small plants and grasses grow on land and hardly any animals live there (apart from breeding birds in summer), but the Antarctic Ocean is teeming with life. The animals have adapted to living in extremely cold conditions, although some, such as whales, leave during the really cold winter, June to August.

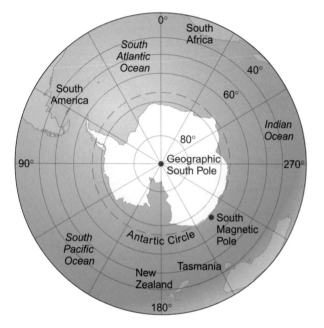

Animals	Adaptation
Whales, seals	Insulating layer of fat
Fish, insects	Chemicals (like anti-freeze)
Penguins, seals	Body shape and thick skin to retain heat
Birds	Waterproof, downy plumage

Animals are all sizes, from tiny crustaceans like krill up to large mammals like whales that eat the krill.

LAND AND SEA ANIMALS

Crustaceans (krill, crabs, shrimps)

Marine invertebrates (squid, octopus, limpets)

Insects (midges)

Fish (cod, skate)

Birds (penguins, albatrosses, petrels, cormorants, gulls, terns)

Mammals (seals, whales, dolphins, porpoises)

The fish, birds and animals form part of a fragile food chain that includes tiny crustaceans called krill and whales that eat the krill. Extensive fishing may damage stocks of fish and krill and then the fragile food chain.

PEOPLE AND ANTARCTICA

The only resident people are research scientists in about 45 bases run by different countries investigating this vast wilderness area (**A**). Great care is taken to avoid damaging the environment and waste is incinerated, packaged up and taken away.

Most tourists visit on cruise liners or on activity holidays and the numbers are increasing very rapidly. There were estimated to be 28,000 tourists in 2004 but if the rate of increase continues there would be 80,000 in 2010.

The greatest impact that people have on Antarctica may be through the atmosphere, through air pollution and global warming. Temperatures are rising in parts of west Antarctica resulting in the ice sheet melting and thinning but in the east the ice sheet appears to be thickening. The warming atmosphere holds more moisture so snow is falling and turning into ice.

A Scientists in Antarctica

Many countries have agreed to try to protect Antarctica and a statement in 1991 designated the area as 'a natural reserve devoted to peace and science'.

Activities

1 Why is Antarctica called a fragile environment?

2 Name ways in which people in Antarctica can damage the environment. Explain how tourists can cause pollution.

3 How may people in the rest of the world damage the Antarctic environment?

4 Draw up a survival guide for people going to Antarctica. Use the internet (see Hotlinks, page ii) and search 'holidays Antarctica' to find details about location, climate, animals and survival. Pick the things you think most important to tell people, what to do and what not to do to protect themselves and Antarctica. You could make up signs to illustrate your message. Try to think of *ten* points.

Ecosystems

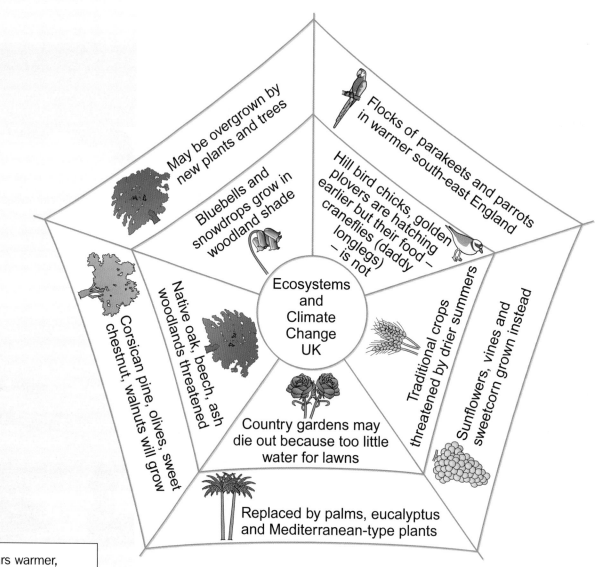

May be overgrown by new plants and trees

Flocks of parakeets and parrots in warmer south-east England

Bluebells and snowdrops grow in woodland shade

Hill bird chicks, golden plovers are hatching earlier but their food – craneflies (daddy longlegs) – is not

Native oak, beech, ash woodlands threatened

Corsican pine, olives, sweet chestnut, walnuts will grow

Ecosystems and Climate Change UK

Traditional crops threatened by drier summers

Sunflowers, vines and sweetcorn grown instead

Country gardens may die out because too little water for lawns

Replaced by palms, eucalyptus and Mediterranean-type plants

Within 50 years warmer, drier summers may make the UK more like the Mediterranean. The climate change report, 2005, assumes summer temperatures in south-east England will be 1.5°C to 3°C warmer than now and summers may be drier. If, as predicted, the major inputs of heat and water alter, ecosystems, from woodland to gardens, will change.

A Web of change

1 Describe the changes shown in each of the sections in source **A**.
2 What could happen to golden plovers if 'daddy longlegs' do not adapt to climate changes at the same rate?
3 Make a list of some insects, birds and animals that you may see in a garden. Who or what do they eat? How might garden ecosystems change if summers become warmer and drier?
4 Does it matter if ecosystems change? Explain why you answer 'yes' or 'no'. Do your friends agree with you?
5 Find out about and explain the changes taking place in a small ecosystem like a garden or pond or in a larger ecosystem such as a desert or forest. What are the causes of the changes? Use the internet (see Hotlinks, page ii) to help you research.

>> 3 Natural resources

The man fishing in the River Ganges is one of millions of people who rely on the river. Is water our most valuable resource?

Learning objectives

What are you going to learn about in this chapter?

> Renewable and non-renewable natural resources
> Changing energy sources
> The need to manage clean water
> How people use and rely on a river
> Who has food
> What should we do with what we do not want?

A Fisherman on the River Ganges at dawn

What are natural resources?

> Finding out about natural resources
> Thinking about how we use them

WHAT ARE NATURAL RESOURCES?

Natural resources are materials that occur in nature and are essential or useful to humans, such as water, air, land, forests, fish, wildlife, topsoil and minerals. They are classified as **non-renewable** or **renewable**. Renewable resources such as fish and forests can be overexploited and stop being renewable, but some are always available, for example solar, tide and wind energy.

A Different types of resources

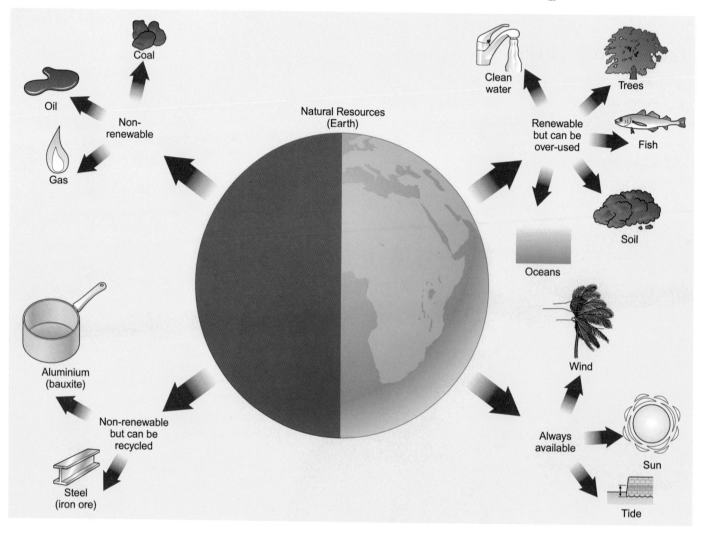

B Recycling symbols

HOW DO WE USE NATURAL RESOURCES?

Some natural resources we use in a lot of different ways. Trees are used to build with, to burn as fuel, to make paper with or simply to look at. We also depend on them to absorb carbon dioxide and give out oxygen so that we have air to breathe. We have built with trees for thousands of years but only recently have we thought of trees as a scenic resource so what we think of as a resource can change over time.

Here are some other examples of how we use different kinds of resources:

- Coal, gas and oil are called **fossil fuels** because they come from the fossil remains of plants and animals. Once burnt, they are gone so are non-renewable resources. Burning them gives us energy but causes pollution. Most of the world's energy comes from fossil fuels.
- Minerals like bauxite can be mined, made into aluminium and, if **recycled**, can be reused time and time again. These are non-renewable, because once dug out they are gone, but they are recyclable. Some countries reuse and recycle much more than others. In the UK we recycle only about 15 per cent of household waste.
- Renewable natural resources such as clean water, oceans, fish, soil and trees should be **sustainable**, there to be used forever, but when treated badly by us they may stop being a resource. Some rivers are so polluted they do not support life. Fish can be over fished until there are insufficient fish stocks to replace themselves.
- Energy from wind, tides and sun is always available, however much we use.

WHO USES MOST RESOURCES?

The more money we have, the better our quality of life and the more resources we use. When you get more money you may buy more things, possibly made from non-renewable resources. When a country has more money (is more developed), more natural resources are used. The USA uses more oil than any other country but China is rapidly developing and in 2004 was the second biggest user of oil. The more resources we use, the more pollution we cause.

MEDCs use the most resources but far more people live in **LEDCs**, use far fewer resources and have a poorer quality of life. Everybody needs food, clean water and some energy at least to cook with, but in parts of the world people do not have even these basics. Natural resources have to be properly managed to benefit people now and in the future.

Activities

1 Copy out this table and complete it.

Resource	Is it non-renewable, renewable, recyclable, always here to use?	How is it used?	Does using the resource cause pollution? If so, what?
Wood			
Coal			
Clean water			
Wind			
Bauxite			

2 List *three* products that can be recycled and reused. How can this 'save' natural resources?

3 Why and how do you think people in the UK use more natural resources than most people in Africa? Which natural resources might be valued most by people in Africa?

4 Design your own diagram to show different types of natural resources. You could include sketches to show how resources are used, for example oil by a car, wind energy by a wind turbine etc. Use other symbols instead of drawings or different shaped words (**C**). For more information you could search the internet (see Hotlinks, page ii) using key words such as 'non-renewable resource', 'coal'.

C Different shaped words to illustrate the meaning

Energy

> **Finding out about energy generation**
> **Thinking about renewable energy sources**

For hundreds of years countries have relied on oil, coal and gas – fossil fuels – for all their energy needs for industry and daily life. However, these are all non-renewable resources (**A**) and we could run out of oil and gas this century. In addition, the ways in which we use fossil fuels causes land and air pollution (**B**).

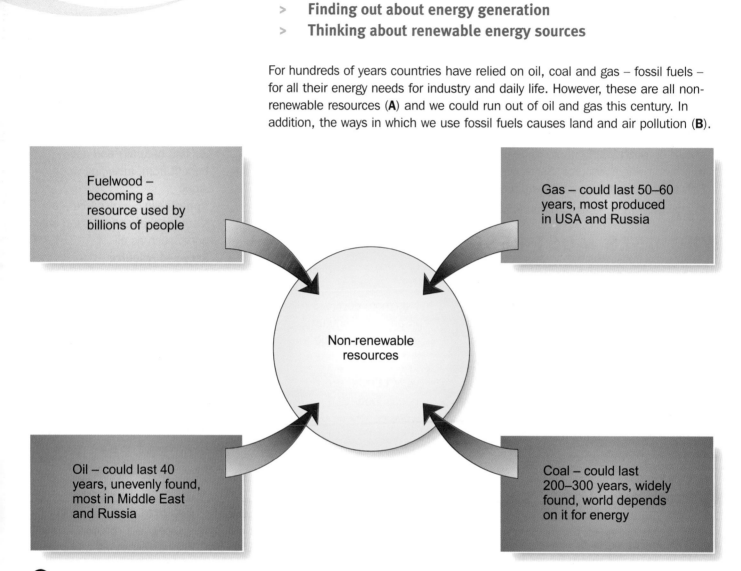

Fuelwood – becoming a resource used by billions of people

Gas – could last 50–60 years, most produced in USA and Russia

Non-renewable resources

Oil – could last 40 years, unevenly found, most in Middle East and Russia

Coal – could last 200–300 years, widely found, world depends on it for energy

A Non-renewable resources

B Power station, Helmstedt, Germany

Renewable **energy sources**, such as water, sun and wind, are cleaner alternatives to fossil fuels. They create little pollution compared with burning fossil fuels but they do have an impact on the environment. It seems likely that most countries will combine the use of renewable energy supplies with more efficient use of fossil fuels and nuclear power.

ENERGY OF THE FUTURE

Ninety per cent of our transport depends on oil: for cars, lorries, planes and trains. Countries use more energy as they get richer and people all over the world want to travel more and use their own cars. People are now developing 'cleaner' ways to travel. By 2030 the world will need 60 per cent more energy than in 2002. Demand will increase in developing countries such as China and in MEDCs such as the USA.

Energy source – something used to provide energy

Fuelwood – any wood that is collected and burnt for fuel, mostly in developing countries

Greenpeace and the Global Wind Energy Council believe that wind can supply 12 per cent of the UK's energy by 2020. Wind turbines (**C**) generate clean energy but views are divided about them. Some people think they have an environmental impact that is unacceptable. Others consider the problems from continued use of fossil fuels are so great that they must be built.

C Wind farm in California

Activities

1. Why do you think more economically developed countries use more energy than less economically developed countries?

2. The world will need more energy in the future. How do you think you use more energy than your grandparents did at your age?

CASE STUDY

Your Energy Ltd has plans for a wind farm in the west of the Isle of Wight with six, 100- and 111-metre tall turbines. The power generated could supply 6,165 homes or 0.8 per cent of electricity needed on the Island. ThWART (The Wight Against Rural Turbines) is a group which considers the proposed village site, near an area of outstanding natural beauty, inappropriate for turbines. Use of energy-saving bulbs in the area might save as much electricity as would be produced by the six turbines.

- Twenty-three per cent of jobs are in tourism, the biggest source of income. Would turbines deter tourists?

- Could tidal energy be developed with less visible impact?

- Is increasing coastal erosion on the Isle of Wight linked with the burning of fossil fuels, global warming and climate change?

3. Read through the case study. Make a list of the points in support of and against a wind farm on the Isle of Wight. Do you think it should go ahead? Justify your answer.

4. Write out definitions of some other renewable energy supplies by matching the beginning of each sentence with the correct ending.

Hydro electric power is generated by water falling	*below the surface of the ground*
Solar power can be converted	*being developed to use the energy in waves*
Wave power equipment is	*onto a turbine, in a dam or down a hillside*
Geothermal power uses heat from	*generated by tidal waters turning turbines*
Tidal power is	*directly to electricity in photovoltaic cells*

5. The yes2wind group of Greenpeace, Friends of the Earth and WWF believe that people should speak out in support of wind power, but others disagree. Check the website of the yes2wind group (see Hotlinks, page ii) for the world's largest offshore wind farm proposal. What is your opinion?

Clean water – the most valuable resource?

> Thinking about water as a resource

> Finding out how people use water

Half the world's population live without access to water that is safe to drink. Why is this when 70 per cent of the earth is covered in water? The answer is that most of this water is seawater, less than 3 per cent is fresh and an even smaller portion is clean and safe to drink. As many as 5,400 children a day die from water-related illnesses.

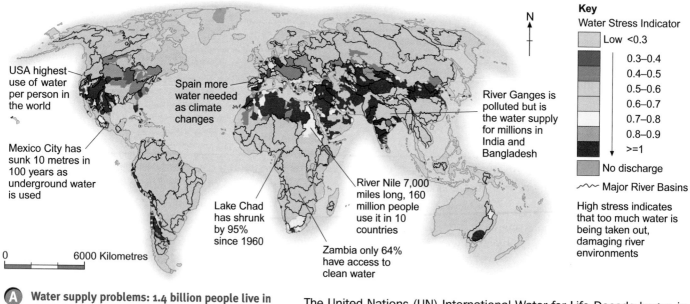

USA highest use of water per person in the world

Spain more water needed as climate changes

Mexico City has sunk 10 metres in 100 years as underground water is used

Lake Chad has shrunk by 95% since 1960

River Ganges is polluted but is the water supply for millions in India and Bangladesh

River Nile 7,000 miles long, 160 million people use it in 10 countries

Zambia only 64% have access to clean water

N

Key
Water Stress Indicator

	Low <0.3
	0.3–0.4
	0.4–0.5
	0.5–0.6
	0.6–0.7
	0.7–0.8
	0.8–0.9
	>=1
	No discharge
～～～	Major River Basins

High stress indicates that too much water is being taken out, damaging river environments

0   6000 Kilometres

A Water supply problems: 1.4 billion people live in damaged river environments
Source: Map courtesy of Washington DC: Water Resources Institute, 2003

The United Nations (UN) International Water for Life Decade began in March 2005. The UN thinks each person needs 50 litres of water a day but billions of people, mostly in LEDCs, have to manage on less. Poor people have the least clean water and spend the most time and money on getting it.

Clean water should be a renewable resource but three things are happening to our clean water supplies. The supply is decreasing, what is left is becoming more polluted and people are arguing over who owns it. Some people are even fighting over water supplies.

HOW IS WATER USED?

Agriculture uses 70 per cent of all fresh water, much of it for **irrigation** (C). Water is easily wasted and one way to reduce irrigation water loss is to put it directly (drip it) into the ground before it can evaporate into the atmosphere. Producing meat uses lots of water (and land as we have seen with rainforest clearing).

- 1 kilo of grain-fed beef uses 15 cubic metres of water (the space you are sitting in could be about a cubic metre).

- 1 kilo of cereal uses only 3 cubic metres of water.

Much less water would be used if we ate more cereals and less meat.

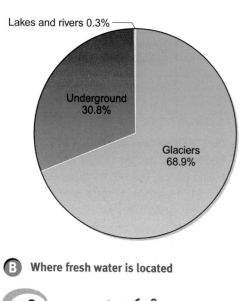

Lakes and rivers 0.3%

Underground 30.8%

Glaciers 68.9%

B Where fresh water is located

The USA

The average American uses 600 litres per day, more than anybody else in the world.

Some 95 per cent of the water used in the USA comes from underground supplies in **aquifers** where water has collected over millions of years, but people are now pumping out more water than is going in and fertilisers and pesticides are polluting supplies. Water is used for all sorts of things from drinking to watering golf courses in deserts, although much of it is used for agriculture. Will people be able to live and farm in ways that use less water?

Spain

The average European uses 250–350 litres a day.

Southern Spain is becoming drier and warmer as the climate changes (20 per cent less rain and 4°C rise in temperature predicted in future), and the demand for water is increasing. Wastewater can be treated and reused but more water is needed. Big schemes can damage the environment so desalination plants are being built to make seawater into fresh, clean water to drink. This helps protect the environment but the desalination plants use oil, a dwindling fuel resource.

Zambia

The average sub-Saharan African uses 35–75 litres a day.

Sixty-four per cent of Zambians have access to clean water. Small-scale, village-size schemes are effective in improving water supplies for the millions of people for whom sanitation and clean water are equally important.

A typical scheme may be a village with 40 households with a village well (simple to maintain with a bucket and winding handle) and 28 latrines (toilets).

Villagers used to collect water from a stream, digging a hole to let water filter through the sand into pools. They were often sick from drinking stream water and from using the bush (the land around the village) as a toilet. Now they have clean water from the well (**D**) and the women and children who used to spend three hours a day collecting water have more time to harvest food or go to school. Diseases like diarrhoea have decreased and villagers, especially children, are healthier. The result is healthier people who have made themselves a better life.

C Intensive irrigation, Libya, North Africa

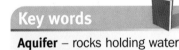

D Village well in Zambia

Key words

Aquifer – rocks holding water underground
Irrigation – artificially watering landscapes

Activities

1 a) Draw a bar graph to show the use of water in the USA, Europe and sub-Saharan Africa. Try to explain what this graph shows. (See page 148 of *SKILLS in geography*.)

   b) Draw a well-labelled diagram or sketch to illustrate the recommended daily water intake.

2 a) Sixty per cent of major cities are on the coast. Are desalination plants a good idea? Explain your answer.

   b) How do *you* think people in the USA could use less water?

3 'Some for all rather than more for some!' What does this article headline mean with reference to water? Write a short article using this headline for your school website.

4 Use the internet (Hotlinks, page ii) to find out more about the water problems in places shown on map **A**. Use the BBC, WaterAid and UN websites or Google. Try to find water projects in developing countries but be specific and search 'clean water Zambia' (or another country).

The River Ganges – a river under pressure

> Thinking about how important the river is to people
> Learning about problems in using the river

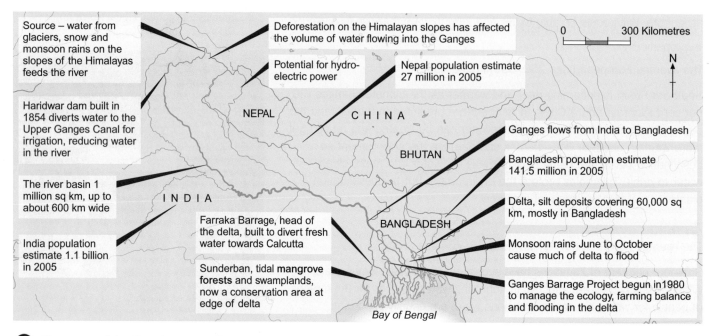

Source – water from glaciers, snow and monsoon rains on the slopes of the Himalayas feeds the river

Deforestation on the Himalayan slopes has affected the volume of water flowing into the Ganges

Potential for hydro-electric power

Nepal population estimate 27 million in 2005

Haridwar dam built in 1854 diverts water to the Upper Ganges Canal for irrigation, reducing water in the river

NEPAL

CHINA

BHUTAN

Ganges flows from India to Bangladesh

Bangladesh population estimate 141.5 million in 2005

The river basin 1 million sq km, up to about 600 km wide

INDIA

Delta, silt deposits covering 60,000 sq km, mostly in Bangladesh

India population estimate 1.1 billion in 2005

Farraka Barrage, head of the delta, built to divert fresh water towards Calcutta

BANGLADESH

Monsoon rains June to October cause much of delta to flood

Sunderban, tidal **mangrove forests** and swamplands, now a conservation area at edge of delta

Ganges Barrage Project begun in1980 to manage the ecology, farming balance and flooding in the delta

0   300 Kilometres

N

Bay of Bengal

A The course of the River Ganges, India

The River Ganges flows from the Himalayas, through Nepal, India and Bangladesh for 2,500 km to the Bay of Bengal (**A**). The river basin is the biggest and most fertile in the world and millions of people depend on the water, but for Hindu people the river is significant as the most sacred River Ganga. Bathing in the Ganges water is believed to wash away your sins (**B**) so millions of Hindu people gather on the banks of the river at festivals. Ashes of the dead are cast into the river to guide souls straight to paradise.

B People bathing in the Ganges

HOW IS IT A RIVER UNDER PRESSURE?

The river is the main source of fresh water, if not clean water, for nearly half the population of India and Bangladesh and nearly all the population of Nepal. These three countries between them have an estimated population of 1.2 billion. The population is increasing but the amount of water flowing into the river remains the same. Water is being taken out (for irrigation, domestic use, industry etc.) and a lot of water is used to produce subsistence food and cash crops such as sugar cane and cotton.

There are also problems with what goes into the river:

- Possibly 1 million litres of untreated sewage flows into the river every day from the towns and cities through which the river flows.
- Chemicals, especially from the leather industry, are a major cause of pollution.
- Partially burnt bodies floating in the river become a health hazard when cremation does not work properly.

The water is used for all domestic tasks even if it is unfit to drink or to wash with (**C**). Thousands bathe in the river daily. More people in India die from water-borne diseases such as diarrhoea and cholera than anything else and many die in Bangladesh when water is contaminated during delta floods. Fewer fish live in the polluted river so more people go hungry. Washing comes out dirtier than it goes in, so people use bleach causing more pollution.

C People washing clothes in the Ganges

OTHER ISSUES

Disputes can occur around rivers that flow through several countries and there are discussions about the Farraka Barrage in India where water is diverted away from the river before it flows into Bangladesh.

The delta regularly floods when the monsoon and cyclone storm rains drive in and meet the Ganges and the Brahmaputra water flowing out. Fresh water flowing out through the delta stops seawater from coming in so if the river flow decreases, more salt comes in and contaminates delta land.

Cleaning the water has become more important as water from wells sunk to find clean drinking water are found to be contaminated with arsenic. People all along the Ganges valley are being slowly poisoned.

Millions of people's lives depend on the Ganges water being successfully managed (**D**). How can this be addressed? In 1985 the Ganga Action Plan (GAP) was set up by the Indian government to try to manage pollution and set up waste treatment facilities. This has only been partially successful and tackling the problems of water volume and pollution is ongoing.

Key words

Mangrove forests – forests which grow around coasts where trees are half submerged in saltwater

Activities

(S)

1. Draw a simple sketch map of the river to show the river (blue) and the country borders (red). Label India, Bangladesh and Nepal, the Himalayas, the delta, the Bay of Bengal and the major cities.

2. Make a Fact File about the River Ganges. Use the information here and add more research using the internet (Hotlinks, page ii) if you can. You could add some photographs.

3. How are people 'putting pressure' on the river – think of *five* problems and explain them.

4. Sketch and label (annotate) one of the photographs (**B**, **C** or **D**) to show how people use the river. (See page 154 of *SKILLS in geography*.)

Food

> Understanding why people are still hungry
> Thinking about food and natural resources

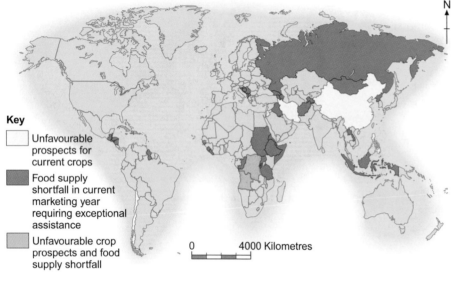

Key

☐ Unfavourable prospects for current crops

■ Food supply shortfall in current marketing year requiring exceptional assistance

▨ Unfavourable crop prospects and food supply shortfall

0   4000 Kilometres

(A) Places in the world where there are food shortages

Source: Food and Agriculture Organization of the United Nations (www.fao.org/documents)

There is more food in the world than ever before, enough to give everybody 2,800 calories each day (500 more than 30 years ago), but we know that people in MEDCs eat a lot more than that: 3,000–4,000 calories a day. The rest of the world gets less (**A**). World food production has increased rapidly but, as with all other resources, the poor are the people who get least.

Every day 25,000 people die from hunger and malnutrition and 300 million people are undernourished in sub-Saharan Africa and 500 million in south and east Asia. Sub-Saharan Africa is one part of the world where there is less food than 30 years ago.

In the UK we can buy any food we want in shops and supermarkets at any time of year – although what you buy depends on how much money you have. So much food is eaten in some developed countries that obesity is a serious problem. The average US citizen eats 122 kg of meat a year and $120 billion dollars a year in the USA is spent on illnesses and problems associated with obesity.

Malnutrition results from a poor diet either with too little food or not the right sorts of carbohydrates, proteins, vitamins or minerals.

People are hungry for lots of different reasons:

- poverty, no money to buy food or seeds
- crop-eating pests, insects and disease
- floods, droughts, mudslides and other hazards
- unreliable rainfall, more frequent with climate change
- wars and fighting stop planting and harvesting of crops
- poor soil, over used, low on nutrients, may be washed away
- crops grown for sale (commercial) leaving too little land to grow food
- illness caused by a poor diet, less ability to farm efficiently.

Some of the things that could help:

- food aid to save lives
- tools, seeds to help people grow food
- more jobs, more opportunities to reduce poverty
- more and better land
- ways to save water
- health care so that families stay well and can work.

(B) Extreme drought, India

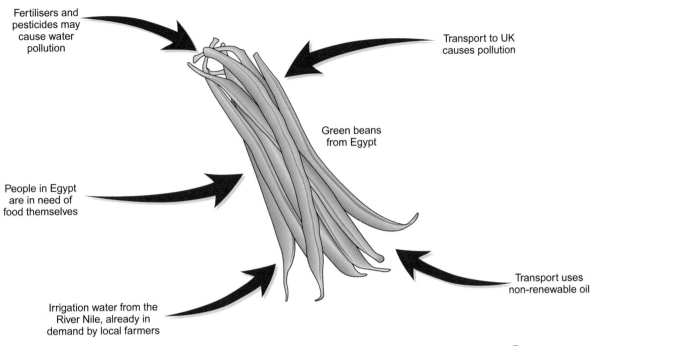

Fertilisers and pesticides may cause water pollution

Transport to UK causes pollution

Green beans from Egypt

People in Egypt are in need of food themselves

Transport uses non-renewable oil

Irrigation water from the River Nile, already in demand by local farmers

Planning food for the future is difficult when rainfall is unreliable and unpredictable. There were 36 countries facing serious food shortages in 2005.

C The hidden costs of green beans grown for us to eat

FOOD OF THE FUTURE

Increasing food production has been possible because of technological changes and intensive use of resources such as water, soil and chemicals. Plants have been developed to increase harvests but many people are worried about using more **GM crops**. Cropland already covers 26 per cent of land and has replaced a third of forests and a quarter of natural grassland. Soil is being degraded (poorer, with fewer nutrients), lost through **soil erosion** and is blowing away in places such as northern China where farmland is turning into desert (**desertification**).

The world population will continue to grow but will we be able to increase food production or should we change how we live and eat more of the things that use fewer resources?

Key words

Desertification – arid land being changed to desert due to drought or over-use
GM crops – Genetically Modified crops, crops whose genes have been scientifically changed
Soil erosion – removal of soil by wind and rainwater

Activities

1 Draw a bar graph to show the following average daily calorie consumption:

Somalia	1,500	Brazil	2,800
Sudan	2,200	UK	3,300
China	2,700	USA	3,700

Draw the bars vertically or horizontally with a scale up to 4,000 calories. Leave a space to add arrows and the following annotations to the graph:

2,800 the world average

2,400 the average needed

1,600 the minimum level for health

**** the calories I eat each day (you may not know this!)

(See page 148 of *SKILLS in geography*.)

2 Subsistence farmers produce three-quarters of world food. Why are so many people still hungry? Explain *three* ways in which people could be helped.

3 Why is obesity a problem? Use the internet (Hotlinks, page ii) and search with key words such as 'calories per person'. Working in groups, draw a mental map or spider diagram and present your findings to the class. (See pages 153 and 155 of *SKILLS in geography*.)

4 How can increasing food production for the future 'put pressure' on natural resources (see source **C**)?

Planet earth – wasted

> Understanding how what we buy affects world resources
> Thinking about waste

HOW TO SPEND £20 BILLION!

Britain throws away £20 billion worth of food a year, an average of £420 per adult. One-fifth of food we buy from supermarkets goes into the rubbish bin unused.

But £20 billion a year, till 2015, would stop 150 million people starving in Africa.

£20 billion is five times the amount we spend on International Aid each year.

Throwing away food is like throwing away natural resources. Every item in the bin represents a resource wasted, such as oil used to produce plastic packaging or water for irrigating crops. Food is discarded when it is past the sell by date but some is thrown away before it gets to the supermarket. Fruit and vegetables thought to be the wrong colour or shape are thrown away or left to rot where they grow.

Whose resources are we wasting? The food we throw away may have been grown in a country where people are paid very little for their work and are struggling to find food for their families to eat. Food left in the fields in the UK wastes our resources.

Waste food and packaging is difficult to dispose of. **Landfill sites** in the UK are nearly full and food is banned from them after 2006. Waste packaging, paper, board, glass, plastic and metal can be recycled but the fastest growing recycling area is composting household and garden waste. Some food waste will be reused successfully but 75 per cent of household waste still ends up in landfill sites. In LEDCs people are so poor they will work in horrible and dangerous conditions on rubbish dumps to find things to recycle or sell (**B**). Children scavenge through rubbish sites to salvage food and yet in MEDCs we throw a lot of it away (**C**).

C Rubbish collection in a wealthy country

A The earth's resources

B Children scavenging on a rubbish dump in a poor country

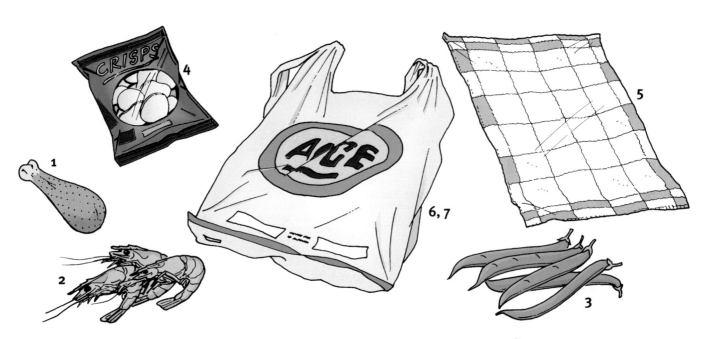

BEHIND THE SCENES

Look at source **D** to see some of the items that make up the global shopping bag.

1 Much of the soya grown in the Amazon, on deforested land, is exported to Europe for animal and chicken feed.

2 Prawns for Europe come from prawn farms in Bangladesh, China, Indonesia and Vietnam. Some coastal mangrove forests have been destroyed to make way for prawn farms.

3 Vegetables grown in Kenya and South Africa are eaten in Europe.

4 Palm oil may be used for crisps, with potatoes grown in the UK. Palm oil comes from plantations on land cleared by burning in places like Indonesia. This causes another problem, smoke air pollution.

5 Considerable resources are needed to grow cotton, in places such as the USA, Russia, Mali, Ethiopia and India. These include fertiliser, 15 per cent of the world's pesticides and water for irrigation. Cotton grown for sale often replaces food crops. The chemicals affect ecosystems and people's health.

6 Oil from the North Sea or the Middle East is used for plastic for packaging and transport.

7 Conifers from Sweden are used for paper and packaging.

Key words

Landfill site – an area where rubbish is collected and dumped then covered over

Activities

1 On an outline map of the world, name the countries producing food and packaging for the global shopping bag. Write in the product used (you could sketch it as well), the resources used and draw a straight line to join the producing country to the UK. Add a title.

2 Which of the countries on your map are LEDCs, lacking sufficient food for all? (You could use the internet via Hotlinks page ii to research this.)

3 How could we use less packaging? Think of food products at home. In pairs, list five things we could do to reduce packaging on our food.

4 Do you waste food? Make a list of any food you have thrown away today or during the week. Why did you throw it away?

5 Definitions of wasted – 'not used to good advantage', 'squandered', 'made uninhabitable'. Do you think planet Earth is being wasted? (Write 100 words or more.)

Natural resources

TOURISTS IN MEDITERRANEAN TOLD TO STOP PLAYING GOLF TO SAVE PLANET

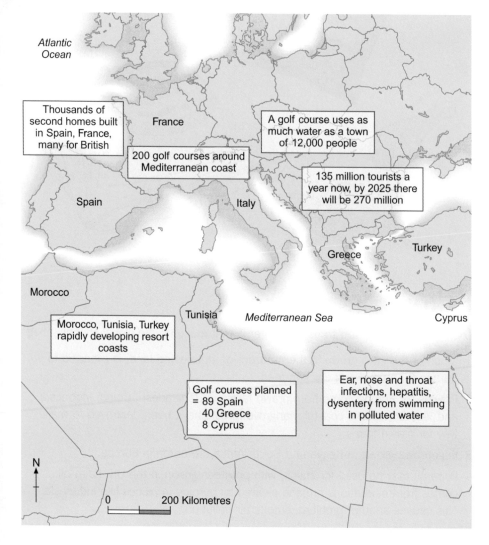

Atlantic Ocean

Thousands of second homes built in Spain, France, many for British

France

A golf course uses as much water as a town of 12,000 people

200 golf courses around Mediterranean coast

135 million tourists a year now, by 2025 there will be 270 million

Spain

Italy

Greece

Turkey

Morocco

Tunisia

Mediterranean Sea

Cyprus

Morocco, Tunisia, Turkey rapidly developing resort coasts

Golf courses planned
= 89 Spain
   40 Greece
    8 Cyprus

Ear, nose and throat infections, hepatitis, dysentery from swimming in polluted water

N

0   200 Kilometres

The World Wide Fund for Nature wrote a report about the Mediterranean region and the effects of tourism. Water supplies may not be sufficient if there are more developments:

- 'The tourism industry depends on water and at the moment is destroying the very resource it needs.'
- 'Mediterranean countries are already raiding limited underground water supplies to keep up with the demand from tourism and agriculture.'
  (An underground water supply is water that has seeped through permeable rock from rainfall and can be pumped up.)

1 What do you think the first bullet point above means?

2 Eco-friendly tourists are urged to boycott golf courses and swimming pools, and take shorter showers. How would this help?

3 What will happen to underground water supplies if more holiday homes are built? How could underground supplies be used up?

4 Tunisia and Morocco are developing countries. Why is tourism so important to them?

5 Suggest ways in which tourism development is putting pressure on water resources around the Mediterranean.

>> 4 Population

Hong Kong is nearly the most crowded place on earth with an average of 6,300 people in every square kilometre. People are living longer. What will this mean for Hong Kong?

Learning objectives

What are you going to learn about in this chapter?

> Population distribution and density

> Population changes and planning for the future

> What happens when people migrate to cities in LEDCs

> The impact of HIV/AIDS on populations

A Hong Kong

Where people live

> Finding out where people live in the world
> Thinking about how population distribution is changing

Key words

Megacities – cities with a population of 10 million people or more

Population density – a measure of how closely people live together, for example the average number of people per square kilometre

Population distribution – how people are spread out over an area

Rural – an area of countryside with small settlements

Urban – a built-up area where a lot of people live

People are not evenly spread out in the world (**A**). We know that the UK **population density** of 240 people per square kilometre is much higher than the European average of 30 people per square kilometre but there are greater variations in other continents. In Mongolia in Asia the average is only two people per square kilometre but in Bangladesh there are about 1,000 people per square kilometre. (Use Hotlinks page ii to look at the US population clock and see the latest world population estimate.)

A Where people live in the world

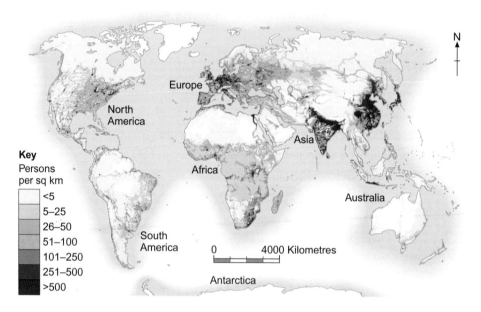

Key
Persons per sq km
- <5
- 5–25
- 26–50
- 51–100
- 101–250
- 251–500
- >500

FACT FILE

World population

1900 = 1.6 billion
2000 = 6.1 billion
2005 = 6.5 billion

Over 3 billion people live in towns and cities but they live on only 2 per cent of the total land available. This means that as the population increases people live in more and more crowded places.

HOW PEOPLE ARE DISTRIBUTED

- The distribution of people over the land is uneven. There are still big, almost empty spaces with very few people because the things we need in life – food, clean water, energy, shelter – are too difficult to find. These are places with extreme climates or extreme slopes (remember the discussion of Antarctica in Chapter 2).

- **Megacities**, big cities with over 10 million people, are new. There were only two in 1970 (New York and Tokyo), but by 2000 there were 18 and 2015 could see 23. Nearly all of them are in developing countries. These cities, such as Lagos, Nigeria and Mumbai, India, are growing in population and the amount of land they cover.

- More people and an increasing proportion of the world population live in lowland places. Millions live in flat, fertile areas like the Ganges Delta, Bangladesh, where there is a high population density, but there is a lower population density in places where the land for farming is poor, such as in northern Quebec, Canada (**B**).

B Area with low density of population, northern Quebec, Canada

- Half the people in the world live within 200 km of the sea but sea levels are rising and some cities are sinking. Bangkok in Thailand sinks 5 cm a year as water is taken out of the ground to use, although it is only 1.5 metres above sea level. We do not really know what will happen to people near coasts as it is difficult to predict how much sea levels will rise (or how much cities will subside).

C Area with high density of population, Shanghai, China

POPULATION DISTRIBUTION IN MEDCS AND LEDCS

The UK is an MEDC. Most people in developed countries live in **urban** areas (**C**). Ninety per cent of you will be looking at this page in school in a town or city.

Until recently, most people in LEDCs lived in villages in **rural** areas but this pattern is changing now as urban areas grow by 1 million people every day. Cities and towns grow because people move into them from the countryside looking for a better life (migration) and through natural increase (more people being born than die).

D Energy usage across the world. Yellow: city lights; red: oil production flares; purple: burning vegetation; green: squid fishing fleets; blue: aurora borealis

Activities

1 a) Plot the following figures as two line graphs to show rural and urban population growth. Put dates along the bottom (x axis) and a scale from 0–5 up the side (y axis). Use two colours and label the lines. Figures are in billions. (See page 147 of *SKILLS in geography*.)

	1950	1960	1970	1980	1990	2000	2010	2020	2030
Urban	0.7	1.1	1.5	1.9	2.2	2.9	3.4	4.0	4.9
Rural	1.9	2.1	2.5	2.8	3.0	3.1	3.3	3.3	3.1

b) What might the total population be in 2030?

c How much bigger will that be than in 2000?

d) How old will you be in 2030?

e) Why is the urban population increasing so rapidly? Give at least two reasons.

2 Look at map **A**.

a) Which continent has few empty spaces?

b) Which continent has the most uneven distribution of people?

3 Look at source **D**.

a) How is the pattern of burning lights different from the population distribution in map **A**?

b) Why do you think there are differences?

4 Look at photographs **B** and **C**.

a) Write a picture in words of the landscape in **B** for people living in a city who may never have been to a sparsely populated place. Write about 100 words.

b) Describe the city in **C** to give a picture of the place in words to people living in rural areas who may never have seen a megacity. Write about 100 words.

Different lives

THE UK, EUROPE, MEDC

Population 60.5 million, July 2005

Median age 38
Land area 241,590 sq km

If you are reading this aged 14 you can expect to live for another 66 years.

As an only child, or with one brother or sister, you will probably live in a small household of three or four people, in a city. Some 23 per cent of you will live with a single parent.

You will expect to have clean water and electricity and 93 per cent of your homes will have central heating, 95 per cent a fridge and 99 per cent a TV.

You may have your own television and computer (54 per cent of homes have a PC) and other possessions. You have to go to school and 50 per cent of you may continue in education after you are 18.

Most of you will work, employed by someone else, until you are 67 (or longer).

As a girl, you may have your first child at 27, get married at 29 (a boy will be 31) and have 1.7 children! One in five girls will not have a child. One in three of you will get divorced at the average age of 40 and probably remarry (two in five marriages are remarriages).

Your parents and grandparents live in their own homes, but you may have to look after them, as you all get older!

All the figures about what life is like in the UK are only averages.

Ⓐ **UK family**

JAPAN, ASIA, MEDC

Population 127 million, July 2005

Median age 43
Land area 374,744 sq km

If you lived in Japan you would live longer, 81 years on average, but in a really small apartment, perhaps in the very crowded city of Tokyo (5,000 people live in every square kilometre). You would spend a long time at school each day and continue education well past the age of 18. Few people would be your age (only 14 per cent are under 14) but a lot would be older, 26 million (20 per cent) are over 65. You might get married at 30 and have one child.

Ⓑ **Japanese family**

 Bangladeshi family

BANGLADESH, ASIA, LEDC

Population 144 million, July 2005 estimate

Median age 22
Land area 133,390 sq km

Living in Bangladesh, one of the poorest, most crowded countries in the world, you would have few possessions of your own but lots of friends as half the people around you will be under 22. Three-quarters of you would live in villages, farming, with parents, maybe uncles and aunts, two or three brothers and sisters but perhaps not grandparents (only 3 per cent of the population is over 65). You may not have clean water or electricity – and your home may flood every year – but you could expect to live till 62. Primary school was from age 6 to 10 but only for 80 per cent of you and at 14 most of you will be working, perhaps in a city, to send money back to your family.

Activities

1. In what ways are you and your family – or your friends – similar to or different from the UK average? Write *seven* questions – for example, 'Do you have central heating?' – to ask your class, family or friends to see how like the UK average you are. You could do a class questionnaire.

2. Look at photograph **C** and the text about family life in Bangladesh. Identify *ten* points that are different from your life.

3. a) Use the government statistics website via Hotlinks (page ii) to find census details. Search 'Living in Britain', a Household Survey, and choose the *five* facts you think most interesting about life in Britain. Explain your choices and compare them with your partner.

   b) Create a class list of facts you did *not* know about the UK.

Key words

Median age – the middle value in a range of ages from youngest to oldest; half the people are younger than this age and half are older

Migration from rural to urban areas in LEDCs

> Finding out about rural–urban migrations
> Thinking about the problems for new city dwellers

The number of people living in urban areas in developing countries is increasing rapidly and one reason is because people are moving to towns and cities from the countryside looking for a better life. Many new city dwellers live in overcrowded slums without adequate clean water and sanitation, but for many from the countryside life is still better in a slum than it was in a village.

WHO MIGRATES TO TOWNS AND CITIES?

Families migrate to towns and cities when they find rural areas too difficult to live in, forced to leave because of hunger or poverty or natural hazards like droughts or earthquakes. These are **push factors**. Young people often move to urban places to find work, perhaps to send money back to family in the countryside, but they also hope to find education, entertainment and healthcare. These are **pull factors**.

PERU

'I work all day with my father on the dump looking for anything to sell. We find the best bits in fresh rubbish but it's dirty work and broken glass cuts me. We came to find work but I would rather be in school.'

KENYA

'We have just arrived from the country because drought killed our crops and our cattle and we had no food. We have built a shelter here on the edge of this shanty town from bits we found but it's the biggest slum in the world and the smell is awful, there are only a few toilets. We do not have enough money to buy the clean water and food we need. I have no work.'

INDIA

'I left my family in the country because there was not enough food for all of us so I came to the city. I live on the platforms and sweep trains and beg for food each day. I get kicked, often, and my friend was injured when he fell off a moving train.'

CHINA

'I am 25 and strong so I can manage my tough city job, labouring on a building site. My family farm, 800 km away and I send them money. I am trying to save enough money to find a girl to marry. I sleep in a hut on site with ten other workers.'

Activities

1. Why do you think each of these people left the countryside?

2. What are the risks for children in cities? Can you think of others not mentioned here?

3. List *ten* problems that new arrivals might experience when they first live in a city. Which would be the worst ones for you? Why?

4. What types of crime might new arrivals face?

5. It is thought that by 2025 60 per cent of people in cities will be under 18 years old (mostly in LEDC cities). Do you think living in a city in 2025 will be fun – or not? Explain your answer.

6. The children in photograph **A** sorting rubbish to recycle can make a tiny bit of money from selling what they collect. Look at the photograph and write what you think could be *their* story of that day.

A Children scavenging in water in the Philippines

Planning for change

> **Thinking how populations change**
> **Understanding the effects of changes**

A Japanese family with several generations

The world population is predicted to be 9 billion by 2050. A lot of things affect how many people live and how long they live but we do know that the populations in many countries are 'ageing'. This means there will be more people over 50 years old than under 50 as younger people in their twenties and thirties grow older but fewer babies are born.

The population has changed in different ways in some African countries. HIV/AIDS has wiped out generations, life expectancy has shrunk to about 35 and population growth has slowed. Predicting the population in 30 years' time is impossible.

Bangladesh
Bangladesh is an example of how populations are changing in some LEDCs in Asia. In the past, families have needed many children to help with work and look after older people but now Bangladeshi families are encouraged to have fewer children and three is the average. In 30 years' time Bangladesh could have an ageing population, like other Asian countries.

Japan
Japan is a fairly typical example of how populations have changed in MEDCs (**A**). People in Japan live a long time but the population will decrease if two children do not grow up to replace two parents. The average woman in Japan has 1.3 children and there is little migration into the country so the population is ageing. Younger people become a smaller proportion of the total. By 2050 there could be 1 million people over the age of 100!

China
Thirty years ago the government in China might not have been able to predict some of the population changes that have occurred. Then China faced a rapidly growing population of nearly 1 billion people and famine seemed a real possibility in the future. Families were restricted to one child only.

Key
- LEDCs
- MEDCs
- Developing

China

India

Japan

Pacific Ocean

Malaysia

Indian Ocean

Indonesia

N

0   2000 Kilometres

B LEDCs and MEDCs in south Asia

1980 ONE CHILD ONLY NOW, CONTRACEPTION COMPULSORY

1995 No brothers or sisters, how will life change?

1983 FAMILIES DESPERATE FOR A BOY!
Traditionally a man and his wife live with his parents and look after them so ....

2005 LIFE EXPECTANCY 72
– will there be enough workers to look after the old?

1985 More girl babies aborted as couples try for a boy!

2004 NOW THERE ARE PERKS TO HAVE GIRLS
– CASH AND PRESENTS GIVEN TO RURAL FAMILIES TO HAVE MORE THAN ONE GIRL

2020 LOOKING FOR A GIRL
– 40 MILLION PERMANENT BACHELORS

2006 One child policy – should it continue?

2005 CARE FOR GIRLS CAMPAIGN GAINING MOMENTUM – FAMILIES GIVEN EXTRA CASH TO LOOK AFTER GIRLS

2000
CHINA FEARS POPULATION CRISIS, 117 BOYS LIVE FOR EVERY 100 GIRLS

2002 GIRLS ABDUCTED AND FORCED TO MARRY, RAPE INCREASES

1982
A rural family needs a boy!
Poor families need a boy to help grow food ...

1986
Girl babies abandoned
– 119 boys live for every 100 girls.

Activities

1 Why did families in China want boys?

2 Sort the headlines into date order and use them to make a timeline of population changes in China since 1980.

   a) Pick out the *five* headlines you think most important and suggest why you have selected them.

   b) How does your list compare with others in the group?

3 Why is the Chinese government encouraging families to value girls?

4 A 4–2–1 family in China is one with four grandparents, two parents and one child. Traditionally families care for each other but do you think this will change in the future? Why?

Population change and HIV/AIDS

> **Understanding some impacts of HIV/AIDS**
> **Thinking about population predictions**

AIDS (acquired immune deficiency syndrome) is a weakening of the immune system by HIV (human immunodeficiency virus). The victim may die from illnesses such as TB, pneumonia, diarrhoea.

When population predictions are made for 10 or 40 years' time, nobody knows what may affect people in the future. HIV/AIDS is causing the greatest changes in population but natural hazards also have a significant impact, such as the 2004 tsunami in south-east Asia that left over 240,000 people dead, many of them children. HIV/AIDS has killed many working-age adults.

All figures are approximate. Accurate data is hard to collect, especially if people do not want to admit to having HIV/AIDS.

By 2005 over 40 million people, including 2.5 million children, were living with HIV. In 2004, 3 million died and 5 million were infected.

AIDS is a pandemic, a global epidemic.

- 70% of cases are in the continent of Africa
- Life expectancy in Botswana has dropped from about 65 to 38 in 15 years
- 25 million Africans are living with HIV/AIDS. A person is infected every 25 seconds
- A baby born in Botswana in 2010 may live for only 27 years
- Over 20 million Africans have died
- Nearly one-fifth of the workforce in South Africa is infected
- Life expectancy has dropped dramatically. Population growth has slowed
- 58% of HIV/AIDS victims in Africa are women so the birth rate in affected countries decreases
- The highest level of infection, at 38% of the population, is in Botswana

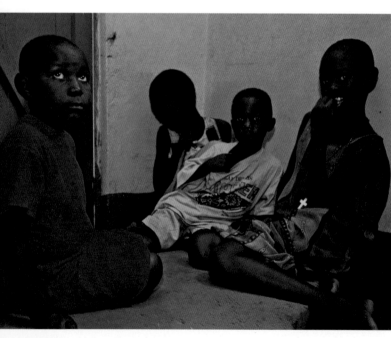

SOME EFFECTS OF HIV/AIDS

- Children become carers and do not go to school
- The loss of adults is affecting all work, including farming
- Less food and money means poorer families, poorer countries
- Poor people are less healthy, more likely to suffer HIV/AIDS, so the downward spiral continues

A Children orphaned by AIDS

WHAT HELPS?

With drugs, HIV/AIDS sufferers live longer with improved health. Many countries just cannot afford the cost of drugs unless they can be produced more cheaply. The government in South Africa, the country with the greatest number of HIV/AIDS sufferers, has pledged to provide antiretroviral drugs to those who need them within five years. People would live longer, have more children and the population structure could change again.

Education is very important (**B**). People need to know how to be safe, how to stop the transmission of the virus and that it is acceptable to use condoms. A more widespread use of condoms does slow the rate of transmission.

WHAT ABOUT OTHER PARTS OF THE WORLD?

AIDS is spreading, especially in Eastern Europe and central Asia. Some governments are trying to slow the spread and the death rate. If more than 1 per cent of the population become infected the epidemic is hard to stop and then population change becomes difficult to predict.

CAN POPULATION CHANGE BE PREDICTED?

Many people living in developing countries are vulnerable to both disease and natural hazards. Who knows what may happen in the future?

Graph **C** is called the African Cliff because of the shape made by the line graphs showing the change in life expectancy in countries hit hard by HIV/AIDS.

B AIDS campaign poster

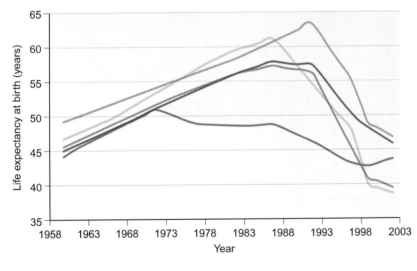

Key

——	South Africa	—— Zimbabwe
——	Botswana	—— Uganda
——	Kenya	

Source: World Development Indicators 2004. Copyright 2004 by World Bank. Reproduced with permission of World Bank in the format textbook via Copyright Clearance Center.

C Life expectancy

Activities

1 Many people in the UK do not know how HIV/AIDS is affecting people and countries. Use the internet via Hotlinks (see page ii) to carry out some research and *either* (a) write a report explaining the facts *or* (b) pick the *five* facts you think most important to tell people about the impact of HIV/AIDS and illustrate them on a poster to be put up in school. You could work with friends and make several posters, perhaps A3 size.

2 Why are education and drugs both important in Africa to help fight HIV/AIDS? Who might provide help for African countries?

3 Describe the changes shown in graph **C**.

Assessing 360°

Population

(All figures are approximate.)

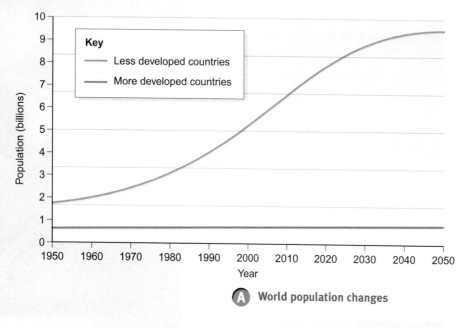

Key
— Less developed countries
— More developed countries

A World population changes

Where people will live

The world population is expected to rise to 9.1 billion by 2050.

1 Describe the changes shown in graph **A** (aim for 50 words).
2 Suggest why people in poor developing countries needed large families.
3 Why do you think a greater proportion of people will live in cities by 2050?
4 Family sizes are smaller now than 50 years ago. How may better healthcare, contraception, AIDS and more money influence the change to smaller families?

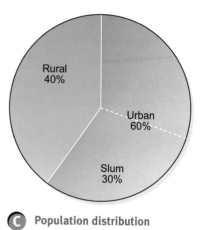

C Population distribution

Rural 40% · Urban 60% · Slum 30%

By 2050 8 out of 10 people will live in developing countries

2 out of 10 people will live in developed countries

B Growth in world population

Rank 2050	Country
1 (2)	India
2 (1)	China
3 (3)	USA
4 (4)	Indonesia
5 (9)	Nigeria

( ) = rank order in 2005

D Ranking of countries by population size

5 Look at source **D**. Why do you think China will be smaller in population than India by 2050?
Which of the largest countries in 2050 are LEDCs now (they may not be in 2050)?
Why are all figures approximate?
6 How old will you be in 2050?

Urban slums

Three billion people will live in slums by 2050.

7 Look at photograph **E** of a shanty town in Indonesia (the fourth biggest country in the world). Which *five* problems do you think are the worst for families living here? Explain your reasons.

E Slum in Jakarta, Indonesia

Cities have a lot of different characteristics and are always changing. That is why they are exciting places to look at but not always easy to understand! Can you think of the advantages and disadvantages of living in a city?

Learning objectives

What are you going to learn about in this chapter?

> How and why cities have changed
> How new opportunities are being created for people in urban areas
> Why crime is a problem in some cities
> How living conditions are being improved in cities
> What is happening on the edge of urban areas

(A) **Run-down inner-city area, London**   (B) **Redeveloped hi-tech inner-city area, Canary Wharf, London**

Changing cities

> Learning about change in urban areas
> Understanding some of the problems faced by urban areas

Key words

Rural–urban fringe – the edge of an urban area where it meets the countryside

Suburb – an area consisting mainly of housing, outside the city centre

A Bullring shopping centre, Birmingham

B Inner city graffiti, Belfast, Northern Ireland

Cities are very complicated places because there are so many different things going on in them. People work and live in cities and also use them for leisure activities. To be able to do all this, cities need to have well-organised transport systems which usually includes buses and trains.

Travelling through any city you notice all sorts of contrasts. There might be fantastic new office buildings, state-of-the-art transport systems and shopping centres (**A**) right next to run-down areas with vandalised buildings (**B**) and streets jammed with traffic.

Urban areas in Britain have developed over a long period of time. Sixty years ago, few people had cars and most people worked close to where they lived, often in crowded inner-city areas of poor-quality housing. Since then cities have continually grown outwards. **Suburban** housing estates, shopping centres and industrial estates have been built on the edges of the urban areas in what is known as the **rural–urban fringe**.

SO WHAT IS THE URBAN CHALLENGE?

Today, many urban areas face particular problems and challenges, some of which can be seen in **C**.

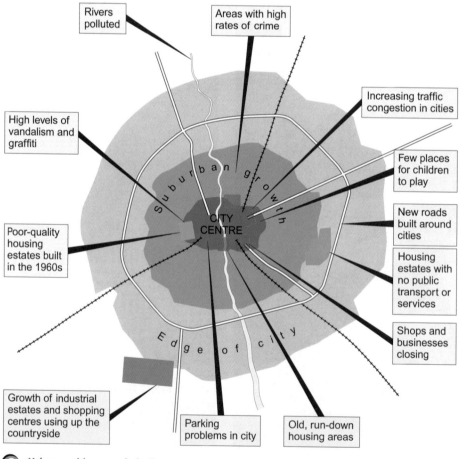

Rivers polluted

Areas with high rates of crime

Increasing traffic congestion in cities

High levels of vandalism and graffiti

Few places for children to play

New roads built around cities

Poor-quality housing estates built in the 1960s

Housing estates with no public transport or services

Shops and businesses closing

Growth of industrial estates and shopping centres using up the countryside

Parking problems in city

Old, run-down housing areas

Suburban growth

CITY CENTRE

Edge of city

C Urban problems and challenges

The cartoon (**D**) shows a meeting of planners discussing the problems of an urban area.

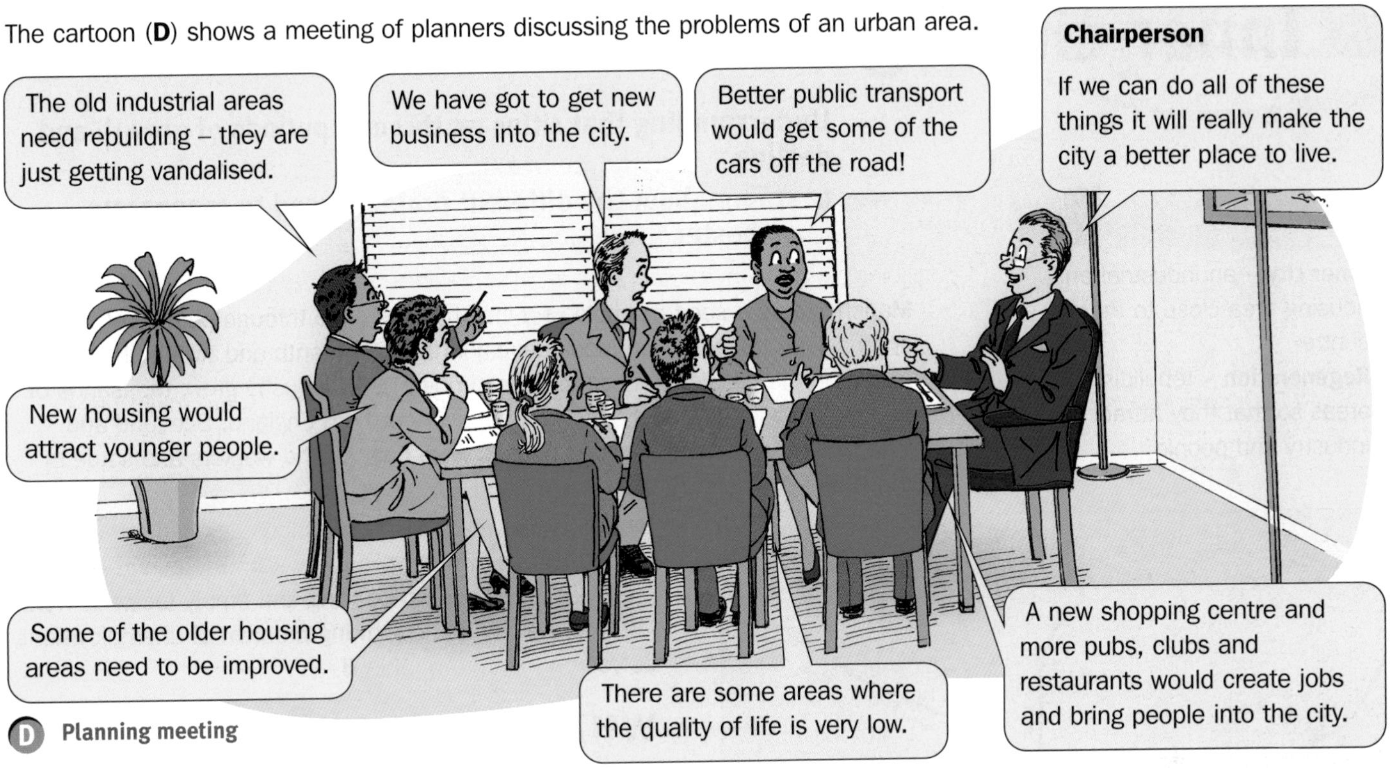

D Planning meeting

WHAT IS MEANT BY 'QUALITY OF LIFE'?

The term 'quality of life' is often used to describe the conditions in which people live. The government compares the quality of life in different parts of cities by using information about a number of factors including those shown in source **E**.

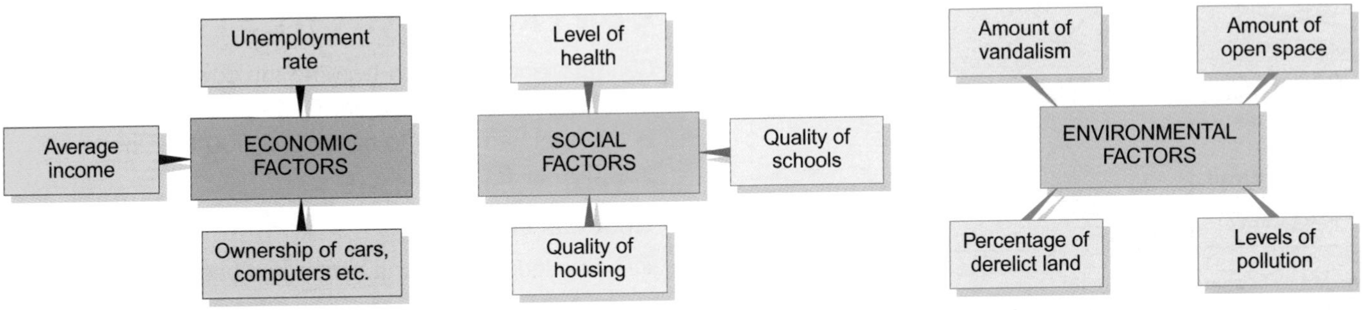

E Quality of life

Activities

1 a) List *five* activities that take place in a city.

b) Why do urban areas need good communication systems?

2 Copy out and complete the table by identifying and explaining another *three* problems faced by urban areas.

What is the problem?	Why is it a problem?
Traffic congestion	Makes it difficult to get to work

3 a) Draw an outline sketch of each of the areas shown on photographs **A**, **B**.

b) Annotate (put notes on) your sketch to identify the main points on each photograph. (See page 154 of *SKILLS in geography*.)

c) What does photograph **B** suggest about the quality of life in this area?

4 Imagine that you are the chairperson of the planning committee in **D**. Write a brief report identifying the main things that could be done to improve the area.

Inner-city regeneration – Manchester

> Understanding that cities go through periods of growth and decline
> Learning about the different projects used to regenerate Manchester

Key words

Inner city – an industrial and housing area close to the city centre

Regeneration – rebuilding areas so that they attract industry and people

Manchester is a good example of a city that has gone through periods of growth and decline. It grew as an industrial city in the nineteenth and twentieth centuries based on the development of manufacturing (**A**). As industry grew, thousands of people moved into Manchester from all over northern England, Scotland and Ireland. Large areas of terraced houses were built for the workers and some of these can still be seen today.

By the 1960s many of the old industries had closed down and people began to move away from the centre of the city. Areas of run-down housing and empty factories were left and the city had a growing problem of unemployment, as you can see from table **B**.

WHAT WAS DONE TO REGENERATE GREATER MANCHESTER?

From the 1980s a number of schemes were developed to help **regenerate** Manchester.

CENTRAL MANCHESTER

Central Manchester had become run down and many businesses had left the area. By 1990 the number of jobs in the area had fallen to below 100,000 from nearly 250,000 in 1950.

What was done?

- A new indoor shopping centre has been built
- Sporting, exhibition and conference facilities have been developed
- Over 3000 houses/flats have been built in the last 10 years

What was the result?

- Thousands of new jobs created
- A big increase in young people living and working in the city
- The city centre is now modern and cleaner

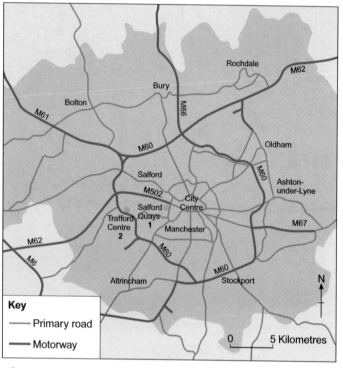

A Map of Greater Manchester

Year	Number of jobs
1950	230,000
1960	160,000
1970	120,000
1980	105,000
1990	95,000
2000	100,000

B Jobs in **inner-city** Manchester

SALFORD QUAYS

(**1** on map **A**) This was part of Manchester Docks which had mainly closed down by the 1970s, leaving empty warehouses and broken-down, vandalised factories.

What was done?

- A hi-tech industrial estate was built
- A number of hotels and leisure facilities were built, including the Lowry art centre
- The old docks have been cleaned up and turned into marinas
- Waterside houses and luxury flats have been built

What was the result?

- Over 4,000 new jobs
- Lots of people now live in the area
- The area is now much cleaner and has landscaped areas with riverside walkways

TRAFFORD PARK

This was an old run-down industrial estate closed down in the 1980s.

What was done?

- A hi-tech industrial estate was built
- The Trafford Centre (**2** on map **A**), a massive shopping and leisure centre was opened in 1998 (**C**)

What was the result?

- Over 50,000 new jobs were created
- The area now has modern buildings and a cleaner environment

METRO-LINK TRAM SYSTEM

The Metro-link tram system opened in 1992 to link housing and industrial areas to the city centre. In 2004 it carried over 19 million passengers (**D**).

C Trafford Centre, Manchester

D Tram in Manchester

Activities

(S) 📄

1 a) What is meant by 'inner-city regeneration'?

b) Why was regeneration needed in Manchester?

c) Draw a line graph to show how the number of jobs in the inner city changed between 1950 and 2000. Remember to label the axes and include a title. (See page 147 of *SKILLS in geography*.)

2 Select *one* regeneration scheme that has created a lot of jobs and *one* that has increased the number of homes. For each one:

– name the scheme

– describe what was done

– explain how the scheme created jobs and improved the area.

3 **Research task** Use Hotlinks (see page ii) to investigate *either* the Trafford Centre *or* the Metro-link. Find out how many people use it, how its use has changed and how it has helped to improve Manchester.

The problem of crime in urban areas

> Understanding what is meant by urban crime
> Learning about the effects of crime on people and businesses

Crime, or the fear of crime, is a growing problem in some urban areas. It can affect the way people live and damage the reputation of local areas.

Burglaries on the increase

PART OF A CITY NOT SAFE AFTER DARK

Vandalism costing local council thousands of pounds a year

Women mugged outside supermarket

CARS SET ON FIRE AFTER FIGHT

A Burnt-out car in a city centre

B Boarded-up businesses in a city centre

WHAT IS CRIME?

A crime is anything that is illegal and punishable by law. Crimes can range from very noticeable actions such as vandalism or criminal damage (**A, B**) to crimes against people such as assault or burglary.

INVESTIGATING CRIME

Students at an inner-city school carried out the following interview with a local crime investigator.

'Is all crime reported?'

No. Most crime is never reported because:

- it might have been carried out by someone people know
- crimes such as vandalism are not against a particular person
- some people think minor crimes are not worth reporting.

'Why are crime rates higher in cities?'

Crime rates are often higher in cities because:

- lots of people live in cities
- there are lots of shops so shoplifting rates are higher
- there are often wealthier areas so burglary rates might be higher
- there is more car crime.

'What are the effects of crime?'

Crime can affect individuals, businesses and local communities:

- *Individuals* – crime causes personal distress/injury and can cost a lot of money.
- *Businesses* – crimes such as shoplifting cost businesses millions of pounds a year. In some cases businesses close down because of the cost of crime.
- *Local communities* – vandalism and criminal damage cost a lot of money to repair. This means less money is then spent on improving local facilities.

'Are people frightened of crime?'

In a recent crime survey carried out in British cities:

- 24 per cent of people said they were worried about burglary
- 18 per cent of people said they were worried about being assaulted
- 30 per cent of car owners said they were worried about damage or theft.

Cities are actually safer than people think – but there is quite a big fear of crime.

'What can be done to reduce crime?'

There are lots of things that can be done to reduce crime. The local crime prevention officer can give good advice which might include:

- fitting security alarms
- fitting better locks on doors and windows
- forming a neighbourhood watch scheme
- improving lighting in outside walkways
- increasing the use of closed-circuit television (CCTV)
- having stricter local regulations on the sale and use of drink.

'Do all areas in a city have the same rates of crime?'

No. Rates of crime vary across the city according to the number of people living in an area, how wealthy areas are, the numbers of businesses in an area and lots of other factors.

Activities

1. a) Identify some of the crimes reported in your local newspaper.

   b) Why might looking at crime reports in newspapers sometimes give a false impression of an area?

2. a) Explain why rates of crime are often higher in cities.

   b) Can you think of parts of your local area where crime rates might be lower or higher than the average?

3. Make a poster with a heading 'What can be done to reduce crime?' Use diagrams and notes to describe what can be done to reduce crime.

4. Draw a bar graph to represent the information under the question 'Are people frightened of crime?'

   - Remember to include a title.
   - Make sure your scale is easy to read. (See page 148 of *SKILLS in geography*.)

5. **Research task** Use Hotlinks (see page ii) to search the internet. Look up one area, town or city.

   a) What types of crime are recorded?

   b) How do the rates of crime vary in your chosen area?

Making inner-city areas better places to live

> Understanding how some areas in cities can become run down
> Finding out how the quality of life can be improved in poor urban areas

Some parts of inner cities have had a number of problems, including crime, poor housing and a lack of job opportunities. This can lead to the gradual decline of an area and make it a very difficult place for people to live (**A**). Source **B** describes what it was like to live in one area of inner-city Manchester.

A Boarded-up house on a Manchester estate

Gangs make inner city a no-go area

BESWICK, an area in inner-city Manchester, is a slum by day, and after dark gangs take over and it becomes a no-go area for the few remaining residents, who hide away behind bolted doors.

You can see skidmarks where stolen cars have been driven over pavements and derelict land to escape police, and burn marks on nearby boarded-up houses show what happened to the cars. The gangs like to race stolen cars in the streets behind the church, now boarded up and awaiting demolition. There are big gaps in the wooden fences alongside the terraced houses, the result of police chases.

Car gangs are only one problem. Lack of jobs is another and there is also drug-related crime. Many people have moved out of the neighbourhood leaving behind shuttered and padlocked homes. Nobody wants to move into the area because it has such a bad reputation. The shopping centre once had a supermarket, a hairdressers, a bank and other shops. Now all that is left is one small shop. The other businesses have closed because of lack of trade and constant problems of security.

B Life in inner-city Manchester

WHAT HAS HAPPENED IN AREAS LIKE BESWICK?

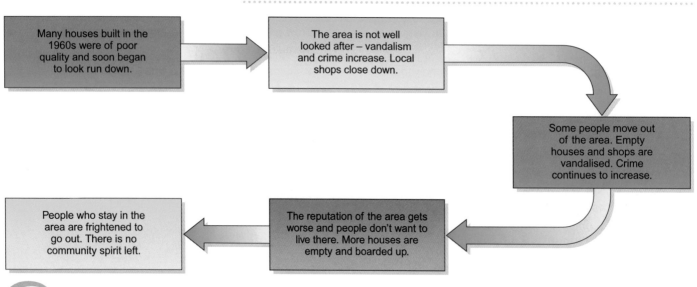

Many houses built in the 1960s were of poor quality and soon began to look run down.

The area is not well looked after – vandalism and crime increase. Local shops close down.

Some people move out of the area. Empty houses and shops are vandalised. Crime continues to increase.

The reputation of the area gets worse and people don't want to live there. More houses are empty and boarded up.

People who stay in the area are frightened to go out. There is no community spirit left.

HOW CAN INNER-CITY RESIDENTIAL AREAS BE IMPROVED?

The following case studies show how two areas in London have been improved.

The Nightingale Estate – Hackney

The Nightingale Estate had many problems including poor-quality housing, lack of safe play areas for children and high rates of vandalism. Part of the area has been renewed to improve the living conditions for local people. Source **C** describes some of the improvements that have been made.

Waltham Forest – east London

Source **D** describes an improvement scheme carried out in east London.

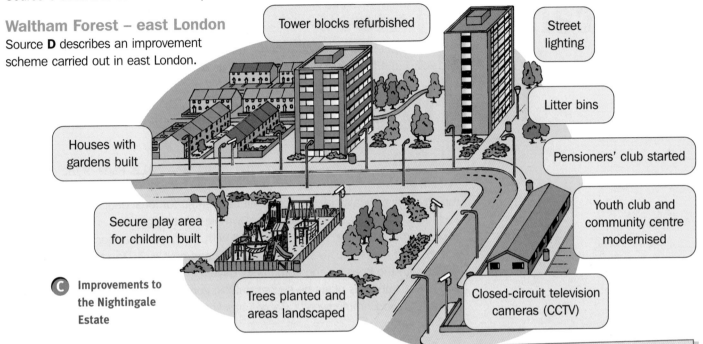

Tower blocks refurbished

Street lighting

Litter bins

Houses with gardens built

Pensioners' club started

Youth club and community centre modernised

Secure play area for children built

C Improvements to the Nightingale Estate

Trees planted and areas landscaped

Closed-circuit television cameras (CCTV)

NEW COMMUNITY RISES FROM RUBBLE OF TOWER BLOCKS

In east London a new community has been formed in what was once a crime-ridden 1990s high-rise estate. This was one of the biggest urban regeneration projects in the country. Five tower blocks were demolished and nearly 1,700 new houses built. Local people have really noticed the difference. One local resident said, 'Nobody used to visit us here because they were frightened of leaving their cars. We didn't like to tell people where we lived because the place had such a bad reputation.

Now friends visit and people are always outside talking to their neighbours or pottering about in their new gardens. There is less crime and the community is much stronger.' The manager of the regeneration project said, 'It's not just about building houses. It's also about building community facilities like schools, health centres and youth clubs and improving people's chances of getting a job. If we can do this, the place has a real future.'

D A new community in east London

Activities

1. Imagine you are living in an area like the one described in source **B**. Write a brief letter to a friend describing what it is like to live there. Make sure you mention:
   - the quality of the houses
   - the general environment
   - the problems of crime and vandalism
   - the lack of shops and services.

2. Why is it very difficult to attract new businesses and shops to run-down inner-city areas?

3. What are the advantages and disadvantages of living in a high-rise flat?

4. Draw a table like the one below and complete it by using information from source **C**.

How the Nightingale Estate was improved

Housing conditions	The environment	Community spirit

Can sport help urban regeneration?

> Thinking about urban regeneration
> Understanding how sport can be part of regeneration

How can  help places like look like

In 2005 London won the bid to host the Olympics in 2012. Parts of the plan include regenerating a huge area of east London, one of the most deprived parts of the city. A high-quality environment will be created for local people and businesses and include a major, new urban park. Olympic sporting facilities will be the focus of new communities, with new housing built with the hope that people's lives will be transformed economically as well as socially and physically. The Olympic stadium will be with the housing and community developments along the Thames estuary. Will these ambitious plans be successful? Regeneration of parts of Manchester as a result of the Commonwealth Games has shown what can be done.

FACT FILE   THE MANCHESTER COMMONWEALTH GAMES 2002

Many parts of Manchester have suffered deprivation and unemployment but East Manchester was particularly hard hit after 30 years' decline. Buildings and land became derelict and decayed, crime increased, house prices fell, poverty grew and people wanted to move away.

The Games project was so attractive that people and businesses invested money in sport, the environment and the people. The Games were part of a long-term plan to improve the East Manchester area with new homes, better transport and amazing buildings and facilities.

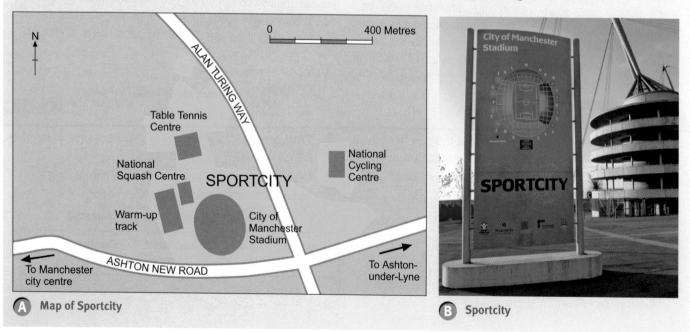

A Map of Sportcity

B Sportcity

Sportcity was the focus of the Games and the plans to make the quality of life better for local people (**A**, **B**). The whole plan will take many years to complete but so far more local people have jobs, more money, facilities to use, new homes and speedier transport to the city centre. There have been 16,000 new jobs (some full time, some part time) created by the Games, in sports, catering, shops and hotels. Over 150 local people have trained as sports coaches.

Some people who volunteered to help with the Games have moved on to other local projects. New, high-quality houses and new, different, affordable housing are helping to make the area a more desirable place to live. There are plans for 12,500 homes.

C The Manchester Stadium

The centrepiece of the £100 million sports complex, the 48,000-seater Stadium (**C**), is now home to Manchester City football club, major sporting events and rock concerts that attract people from all over the world, as well as from Manchester and north-west England.

D Manchester at night

Every visitor spends money and 300,000 more visitors a year now come to Manchester (**D**). A global audience of one billion people saw Manchester during the Games.

Were the plans successful for all?

Not everyone feels the same about the benefits from the Commonwealth Games. Some people thought that repairing a local facility such as a swimming pool would be of more benefit to people than building a large swimming pool to serve a region, but pride in the city has grown and New East Manchester has clearly changed.

More of this £ and a lot of 🏟
mean more of 👥

Activities

1 Why did East Manchester need regeneration?

2 Look at **A** and **B**. What facilities were built for the Games? Planning for their future use has made the regeneration successful. Who uses the facilities now and in what way? You could draw a diagram or map.

3 Choose *five* changes that you think have helped local people most. You could make up your own posters to illustrate the changes.

4 There are conflicting opinions about whether the Games have been good for Manchester.

| Games were a bad thing for Manchester | Games were a good thing for Manchester |

a) Where do you think families, businesses and shopkeepers would fit on the line between the two opposite opinions?

b) Where would you fit?

c) Write a paragraph explaining your answer.

The rural–urban fringe

> Finding out what is in the rural–urban fringe
> Understanding why people disagree about how the land is used

The rural–urban fringe is land around the edge of an urban area that is constantly changing as urban limits are redefined with each new urban development plan. People from both town and country want to use the land but sometimes there is disagreement or conflict about how it should be used.

Towns and cities expand as businesses need more space and spread outwards into the rural–urban fringe and the urban population increases. This expansion is called **urban sprawl**.

The rural–urban fringe is an attractive area to live in and some people believe it should be protected from too much development. Some cities have land around them identified as a **green belt** – land that should be left as rural (**A**).

More families and individuals want their own homes. The easiest place to build new homes is a field on the edge of a built-up area, called a **greenfield site**.

City centres are congested, so anything new that covers a lot of land – such as leisure centres, warehouses, park and ride car parks, roads, bypasses and shopping centres (new shopping complexes have to be built inside cities now) – is easier to build on the edge. Countryside is built over and the rural–urban fringe is built up and changed.

Key words

Green belt – land around cities where building is restricted
Greenfield site – land on the edge of a built-up area
Urban sprawl – expansion of an urban area into the rural–urban fringe

A Green belt

People from homes outside the green belt drive to city

GREEN BELT

Urban built-up area

Ring roads, motorways through green belt

B The rural–urban divide

Look after birds and wildlife.

We want the countryside protected.

We want jobs.

I have always lived here.

I want to farm but also to sell some land.

We retired here. We don't like change.

What do people in the countryside think about the use of land in the rural–urban fringe?

- Local people and retired people may not like change.
- The Royal Society for the Protection of Birds (RSPB) and others concerned about the environment may want to protect landscapes and habitats.
- Farmers want to farm but may want to build on some land.
- Young people want jobs.
- Developers want to build.

How do people in towns and cities want to use the edge of the countryside? Possibly for ...

- New houses in peaceful, safe places.
- Leisure and tourism activities such as golf, walking, riding, etc.

- To drive through or park in (park and ride car parks).
- Warehouses, factories.

Who decides what the land is used for? Should it be ...

- The government who say we need more homes.
- The local council who manage development in town and country.
- Farmers who may want to change how they use their land.
- Village people who do not want to become part of a town.
- Town people (90 per cent of us) who want to use the countryside.

Activities

1 Write your own definitions for:
   - rural–urban fringe
   - greenfield site
   - green belt.

2 Suggest *three* reasons why it is easier to build on the edge of a city than in the centre. (Remember the cost of land in town and country.)

3 Should landowners be able to use the land as they want? Explain your thinking.

4 Which countryside people may have conflicting ideas about how the land should be used? Copy out and complete the table on the right, adding some more groups if possible.

Group	Opinion
Developers	Want to build on the land
Farmers	
Retired people	
Walkers	
Young people	

5 Design a spider diagram to show how town people use the rural–urban fringe. (See page 155 of *SKILLS in geography*.)

6 Who do you think should decide how land in the rural–urban fringe is used? Do you think others in your group agree with you? You might like to have a class discussion about this.

Urban change

The last 50 years have seen tremendous changes within the rural–urban fringe areas in the United Kingdom and these changes are likely to continue as demand for building land increases. It has been suggested that nearly four million extra homes will be required by 2021, with half of them being built on land on the edge of towns and cities. Along with houses comes the demand for other forms of development, such as roads and shopping centres.

1 a) What is meant by 'the rural–urban fringe'?
   b) A lot of building is taking place on the edges of towns and cities. Describe some of the different types of development taking place in these areas.
   c) How might building in these areas damage the environment?

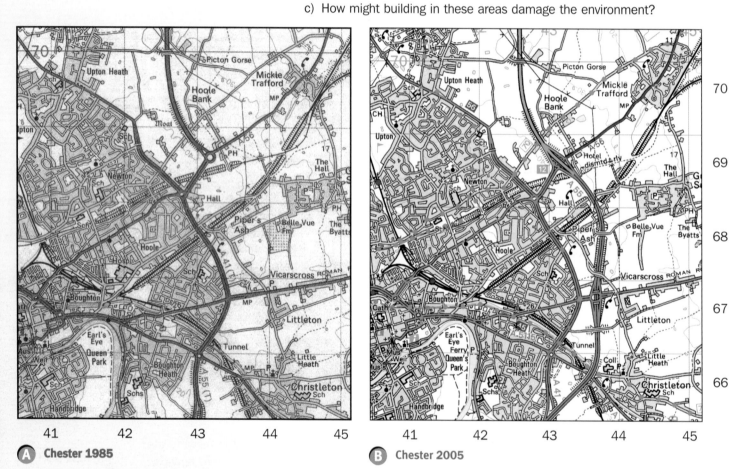

A Chester 1985

B Chester 2005

2 Study the maps in sources **A** and **B** which show the eastern edge of the city of Chester in 1985 and 2005.
   a) Describe how the area shown on the maps has changed between 1985 and 2005.
   b) (i) Suggest why Mickle Trafford has grown.
      (ii) Why might some of the residents in Mickle Trafford have different opinions about the growth of the village?
   c) Why are housing developers keen to build in areas on the edge of towns and cities?

(See page 144 for Ordnance Survey map symbols.)

More than one billion people live on less than $1 a day in extreme poverty. How can we enable everyone to have a better quality of life without permanent harmful change to the planet?

Learning objectives

What are you going to learn about in this chapter?

> The quality of life in different countries
> How aid and trade can influence lives
> How development has damaged the land environment and the atmosphere
> Why global warming and climate change are happening
> Clean and appropriate energy
> Sustainability and the Millennium Development Goals

 **São Paulo, Brazil**

People and development

> Finding out what development means for people
> Thinking about quality of life

Development is about improving the quality of life for individuals so that all people can live safely with the basic things they need such as food, housing, work, education and healthcare. Some countries are more developed than others but within a country each person may have a different quality of life. Clean running water from a tap is a basic necessity for all people in a developed country but only for a few in many developing countries.

The family in Sudan, Africa (**A**) live in a remote place with few possessions where the task each day is to find the basic things they need to survive. Without communications such as telephones or televisions it is hard for them to know how other people in the world live. The things we take for granted such as schools or a bar of chocolate are unknown to them. With very little healthcare, how long can people in this family expect to live? How will children survive simple illnesses such as diarrhoea? What would improve their lives?

0   2000 Kilometres

SUDAN, AFRICA – WHAT IS THE COUNTRY LIKE?

Population 40.2 million

Median age 18

Life expectancy at birth 58 years

Infant mortality 65 per 1,000 live births

Literacy – 61% aged over 15 able to read and write

(A) Family in Sudan, Africa

The people in Ecuador (**B**) do know how other people live. Like thousands of people, they live in homes they have built themselves and perhaps are able to see the world on satellite TV using electricity supplies that may be illegal and dangerous. There are shops, a school and some healthcare but the family quality of life depends on having enough money to buy what is needed. Work is often hard to find so everybody in the family has to try to help.

ECUADOR, SOUTH AMERICA – WHAT IS THE COUNTRY LIKE?

Population 13.4 million

Median age 23

Life expectancy at birth 76 years

Infant mortality 24 per 1,000 live births

Literacy – 93% over 15 years of age able to read and write

0   2000 Kilometres

(B) Family in Ecuador, South America

THE USA, NORTH AMERICA – WHAT IS THE COUNTRY LIKE?

Population 296 million

Median age 36

Life expectancy 78 years

Infant mortality 6 per 1,000 live births

Literacy – 97% over age 15 able to read and write

0 ____ 2000 Kilometres

C Family in the USA

The family in the USA (**C**) live a life that depends on the work that they do and the society they live in. The size of their home, their possessions, holidays, etc. depend on how much money they can earn, but the healthcare and education they use is available to all people in their developed country. They expect to have electricity and clean running water, and pay for it, but the more money they have to spend the more choice the family has in where and how they live.

Key words

Development indicator – information to find the level of development of a country, such as infant mortality rates

Infant mortality – number of babies in every thousand born alive that die in the first year

Life expectancy – average lifespan at the time of birth

Activities

1 Look at the photographs (**A, B, C**).

a) What family possessions can you see in the photograph of the family in Sudan (**A**)?

b) What is the neighbourhood like in Ecuador (**B**)? Describe the general appearance of the building and the immediate area.

c) Find *three* things in the photograph that give you information about living in the USA (**C**).

2 Use the **development indicators** in the boxes above.

a) Describe the differences in life expectancy between Sudan, Ecuador and the USA.

b) Which country has the poorest healthcare?

c) The infant mortality rate indicates how many babies get good healthcare in their first year. With good healthcare, the infant mortality rate is low. Which two countries have high infant mortality rates?

d) Which country has the lowest literacy rate and the highest proportion of young people? (Look at the median age.)

3 Use information from the photographs (**A, B, C**) and the development indicators in your answer.

a) Which do you think is the poorest, least developed country?

b) Which is the wealthiest, most economically developed country?

c) Do you think Ecuador is an LEDC or an MEDC?

4 Which of the following could be most helpful to you in deciding if a country is developed, becoming developed or less developed:

– number of people per doctor

– number of internet users

– average daily calories intake per person?

Development and work

> **Looking at people and work**

> **Thinking about changes in work**

The wealth in a country and the way people work are clues for deciding how developed a country is. The wealthier the country, the more money may be spent on improving things such as healthcare and education for all. People in a wealthy country are likely to have different types of jobs from people in a poorer country.

GDP

The wealth produced in a country in a year is often shown as the Gross Domestic Product or GDP per person. This is calculated by dividing the total wealth in dollars produced in the country in a year by the number of people in the country. GDP is only a guide and does *not* mean that everybody has an equal amount of money each year.

Poverty means not having the necessities to live so the number of people in poverty in each country will depend on the cost of living (see Fact File).

TYPES OF WORK

Primary, secondary and tertiary are names given to types of jobs. Primary workers in farming, forestry, fishing or mining generally earn less money than those doing secondary jobs, making products to sell, or tertiary jobs, providing a service.

A Primary worker

B Secondary worker

C Tertiary worker

Primary workers

Subsistence farmers who grow their own food (as in Sudan, **A**) are generally poor with little money to buy food if a disaster affects their crops. Other primary workers grow cash crops to sell. For example, flower growers in Ecuador earn money from selling their produce but only as long as richer, more developed countries will buy the flowers.

Secondary workers

The wage paid to a worker making a product is often the biggest part of the price of that product. Manufacturing companies may employ people in different countries, where they can pay the least amount to a worker skilled to do the job (**B**). One reason very poor countries have few skilled workers compared with more developed countries is because they cannot afford to educate and train them.

Tertiary workers

Human resources can be more than physical labour such as that used in farming or mining. People use their skills to provide services in a huge range of tertiary jobs from nursing to call centres (**C**).

Informal work

Many people in the world work in the 'informal sector'. They make their own way, doing whatever they can, e.g. building, running a bar, making matches, pots, jewellery, etc.

Global communications such as satellites and the internet enable companies to employ workers in one country to provide a service for people in another country. For example, people in Indian call centres work for Australian and UK banks (called business processing outsourcing – BPO). Skilled operators in India are paid less than similar operators in the UK but the money is sufficient for a good quality of life in India (about £150 per month in 2005). UK operators would have a poor quality of life in the UK if they were paid the same money as people in India.

Country	GDP per person	People in poverty
Sudan	$2,100	40%
Ecuador	$3,900	45%
USA	$41,800	12%

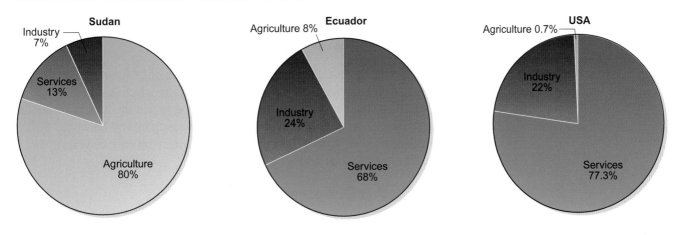

D Pie graphs showing distribution of work force in Sudan, Ecuador and the USA

Activities

1 Look at the figures for GDP for Sudan, Ecuador and the USA in the table above. Explain which country may have the best healthcare and education for all and which may have the poorest.

2 a) Copy out and complete the following table by putting the jobs in the box below in the correct spaces.

Type of job			
Primary			
Secondary			
Tertiary			

rice grower	shop assistant	holiday rep
T-shirt maker	tree logger	table maker
call centre worker	fisherman	car maker

b) Write down *five* other tertiary jobs.

c) What sort of job do you want to do?

d) In which types of jobs do people in your family work?

3 a) Which of these is an *informal* worker in London:
   – a street entertainer?
   – a bus driver?

b) Is a call centre job formal or informal employment?

c) Why do you think there will be a lot of people in Sudan working 'informally'?

4 Use the information on employment, GDP and poverty above to complete the following sentences:

The USA is an MEDC because ...

Sudan is an LEDC because ...

Ecuador is an _____ because ...

Trade and development

Key words

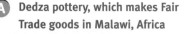

High value goods – products that sell for high prices such as computers

Low value goods – things that sell for low prices such as bananas

Transnational corporations – massive global businesses such as Ford, McDonalds or Nike

Trade is often seen as a way for poorer countries to improve levels of development. Many LEDCs have relied on selling raw materials such as timber, food crops or minerals but development now depends on working people being an important part of trade.

Some developing countries trade in manufactured goods because they have a large workforce of people who work for low wages and they can attract big companies such as Nike to set up factories. Money coming into the country from international companies (called **transnational corporations**, TNCs) can help development but if wages are lower in another country the company may move work there.

These transnational corporations often have their headquarters in a developed country such as the USA, so much of the money from trade still goes to MEDCs although the factories are in LEDCs.

Vietnam is aiming to develop rapidly, following the lead of China in attracting many TNCs. Nike employs 130,000 workers in Vietnam, which is a throughput country for the American firm – raw materials coming in and manufactured goods going out made by workers who will be paid much less than workers in the USA. The literacy rate is very high at 94 per cent for a developing country where 60 per cent of people still work in agriculture.

FAIR TRADE

Poor farmers and workers find it difficult to get a fair deal in global trading, so fair trading aims to help them get fair wages and safe working conditions in return for a good quality product. Fair trade labelling tells the consumer that the standard of the product is good and a fair price has been paid to the producer. Traidcraft is an organisation that buys from poor people at a fair price and so helps them to build better lives through fair trade (**A**) (see Hotlinks, page ii).

A body called Fair Trade Labelling Organisations International (FLO) is responsible for overseeing fair trade, which allows more than 800,000 producers and workers and those who depend on them in 50 countries to benefit from use of the Fair Trade label (see Hotlinks, page ii).

A Dedza pottery, which makes Fair Trade goods in Malawi, Africa

FLOWER WORKERS

Most cut flowers sold in the USA come from Ecuador and Colombia (**B**). About 15 million stems a year are cut and packed and two-thirds are flown to the USA within 24 hours. Many of the 100,000 flower workers are women and children, paid a low wage, $145 dollars a month is typical, working long hours, bending, breathing pesticides and being cut by thorns. The wage may be the only income in the family. More hours are worked before Valentine's Day.

B Rose production in Ecuador, South America

Activities

1 Look at the following table.

	Some exports	Some imports
Ecuador	Petroleum, bananas, flowers, shrimps	Vehicles, medicinal products, electricity, telecommunication products
USA	Motor vehicles, computers, telecommunications equipment, aircraft, medicines	Raw materials – agricultural products, petroleum, manufactured goods

a) Which country exports mainly low-value raw materials?

b) Are the imports to Ecuador **high value goods** or **low value goods**?

c) Does the USA export goods of high or low value?

d) Which country do you think will make most money from its export trade?

2 a) What are conditions like for some flower workers in Ecuador?

b) A rose that sold for $8 in the USA cost 20 cents to produce in Ecuador. Who do you think made the money?

3 Copy the diagrams on the right and put the labels in the box below in the correct circles. You could sketch examples of products from the table above.

Developed world   Developing world

Import low-value raw materials
Export high-value manufactured goods
Getting richer and more developed
Export low-value raw materials
Import expensive manufactured goods
Getting poorer

4 a) How has fair trading helped improve the quality of life for many workers and their families?

b) If two jars of honey looked similar, which one would you buy – a fair trade one or a slightly cheaper one? Why?

5 Why will Nike prefer to run factories in countries with a high literacy rate?

6 **Research task** Use your local shops and the internet (see Hotlinks, page ii) to investigate:

– how roses are produced and sold in Ecuador

– products with the Fair Trade mark and the countries they come from: for *one* craft and *one* food product, find out who produced them, and where and how Fair Trade has improved the quality of life for the producer.

Aid and development

> Finding out about aid
> Looking at how aid helps small-scale development

Aid is help given to people in need. It can be food, money, technology, medicines, new roads, education, whatever is needed.

WHEN IS IMMEDIATE OR SHORT-TERM AID NEEDED?

Immediate aid is needed in a disaster or conflict to rescue people and save lives but the type of aid will depend on the situation. Medicines are always needed but helicopters may be most useful when roads are destroyed or clean bottled water may be more useful in flooded areas. Damage to the infrastructure in the Pakistan earthquake in 2005 was considerable and aid from other countries was vital to get help to isolated places (**A**).

HOW CAN LONG-TERM AID HELP DEVELOPMENT?

Aid that helps people improve the quality of their lives will help development. Aid may come from governments of developed countries, the United Nations, large international organisations such as Oxfam, Save the Children Fund, WaterAid or smaller groups or charities. Small changes can bring development to a lot of people.

Source: The case studies on pages 90 and 91 are adapted by the publisher from the websites http://www.oxfam.org.au/world/africa/ sudan/microcreditidp.pdf and http://www.oxfam.org.uk/what_you_can_do/ give_to_oxfam/company/cow_loan with the permission of Oxfam GB, Oxfam House, John Smith Drive, Cowley, Oxford OX4 2JY, UK. Oxfam GB does not necessarily endorse any text or activities that accompany the materials, nor has it approved the adapted text.

CASE STUDY
OXFAM MICRO CREDIT PROGRAMME, SUDAN, AFRICA

Hassanat and Christina in Port Sudan are busy working themselves and their families out of poverty with the help of Oxfam and a micro credit programme. Both women were displaced by war and have families to look after. The micro credit programme allows each group member to take out a small loan, about $100, to start a business. Hassanat and Christina bought flour and oil and began making pasta and biscuits to sell. In ten months the loan was paid but a bigger loan provided a refrigerator and they made ice-cream to sell. The money enabled them to build concrete houses, pay school fees for all the children, buy medicines, some furniture and they also went to adult literacy lessons themselves. Everybody benefited.

OXFAM COW LOAN SCHEME, MALAWI, AFRICA

When drought or bad harvest destroy the crops that farmers in Malawi rely on, poor families can become destitute. Oxfam has worked with a local organisation, Shire Highlands Milk Producing Association, to provide help through the cow loan scheme (**B**).

Families in need are each given a cow by the scheme. A cow may produce as much as eight litres of milk a day. The family keeps some of this and the rest is sold. When the cow gives birth, if the calf is female it is given back to the scheme to repay the loan. When this calf grows up it is loaned to another family in need, and when it in turn has a calf, this calf is returned to the scheme, and so on. If the calf is male, it is kept by the family and reared to sell. Thanks to the scheme, poor families can afford food, soap and other basic needs. The loan of a cow has transformed their lives.

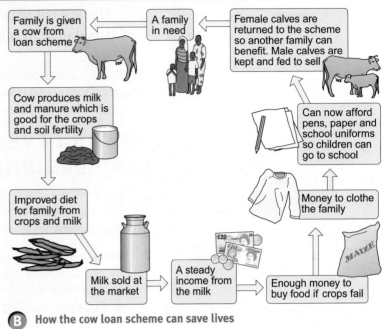

B How the cow loan scheme can save lives

Activities

S **A**

1 a) Look at the photograph showing aid in Pakistan after the earthquake (**A**). Make a list of *ten* aid items that would help these people immediately, e.g. tents for shelter, bottled water.

   b) Why might building new houses be a good long-term aid project?

2 Aid has to be useful to all the people affected. Here are two different aid situations.

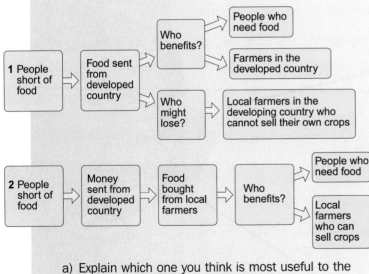

   a) Explain which one you think is most useful to the people in the country needing the aid.

   b) Who do you think benefits most from the second type of aid?

3 Look at source **B**. Explain how the process of development works with just a cow. You could sketch your own large flow diagram to show all the people who may benefit, including the people who buy the milk and the people who sell to them.

4 How can a loan of just $100 help families build houses, pay school fees, buy medicines and build a better future? Write about 100 words.

You could try to assess your own working level by matching your work with some geography level descriptions, or you could use the following levels as guidelines for your writing.

Do you think your writing shows that you:

- understand how the loan leads to changes that affect the lives of people (Level 4)

- have described and explained how the loan can lead to changes and how the changes are linked (Level 5)

- have shown understanding of the sequence of development (Level 6)

- have used all the above and shown an understanding of people's values and attitudes (Level 7)?

Development and the land environment

> **Finding out how development changes the environment**

> **Looking at development and waste**

Development seems to mean more of everything: more houses, more computers, more telephones, and even more garden furniture. Quite often, having more changes the land environment and generates more to throw away.

HOW IS DEVELOPMENT CHANGING THE LAND ENVIRONMENT?

1 **Land** We are changing the land environment with increasing numbers of buildings, roads, shops, factories or anything that is part of a built environment.

2 **Using resources** Taking resources out of the ground to use – trees, minerals, rocks or fossil fuels – changes the environment. Forests, both tropical hardwoods and coniferous soft woods, have been cleared. Mines and quarries have left enormous holes in the earth surface.

3 **Pollution** Our environment changes when we pollute it. Oil leaks devastate parts of the countryside where oil is extracted. Landfill sites (rubbish dumps) cover land near every town. We are still not absolutely sure how to get rid of toxic and nuclear waste without making the land dangerous to people.

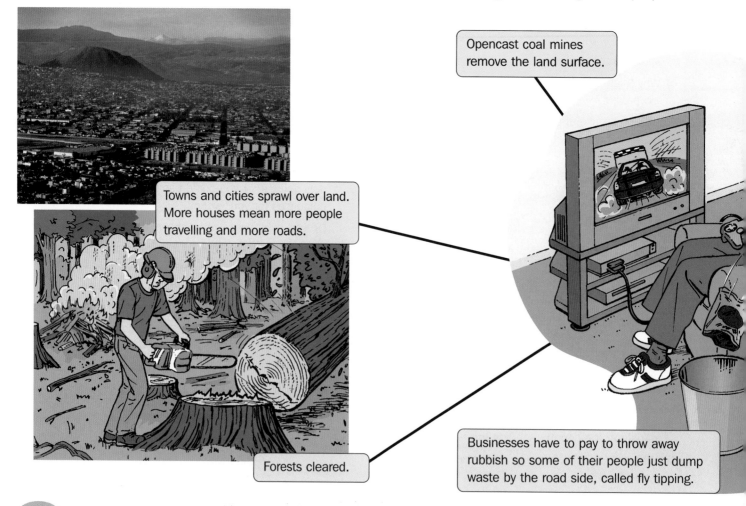

Opencast coal mines remove the land surface.

Towns and cities sprawl over land. More houses mean more people travelling and more roads.

Forests cleared.

Businesses have to pay to throw away rubbish so some of their people just dump waste by the road side, called fly tipping.

1 Do you think it matters that more land is being covered by buildings and roads? Give your reasons.

2 Iron ore is used to make steel. How can we reduce the amount of iron ore we use?

3 Think of *three* products in your home that could contain chemicals that may damage the environment when thrown away, e.g. gloss paint.

4 a) Draw a bar graph using the information below to show what materials were thrown away in the USA in 2003. (See page 148 of *SKILLS in geography*.)

- Paper 35%
- Yard trimmings 12% (called garden refuse in the UK)
- Food scraps 12%
- Plastics 11%
- Metals 8%
- Textiles, rubber, leather 8%
- Glass 5%
- Wood 6%
- Other 3%

b) Add up the percentages for the first *four* items above. Do we throw away similar bulky types of waste in the UK?

c) Which things could be burnt in an incinerator?

d) Which of these things can be recycled?

5 Find out about the waste in school, e.g. electricity use, packaging, food, etc. In groups, decide how you will search for information. You might use surveys, questionnaires, interviews. Draw up a plan to reduce the waste. You could present it to your headteacher.

Old fridges can only be broken up inside special buildings because they contain greenhouse gas chemicals.

Waste disposal is increasingly difficult and expensive in the developed world. We have created 'mountains' – 'waste mountains', 'tyre mountains', 'fridge mountains'.

Hazardous waste is poisonous and should be disposed of safely but increasingly land is contaminated by toxic and nuclear waste.

Much waste from homes is packaging – paper, cardboard and plastics.

A How much pollution results from our comfortable life?

The atmosphere and development

> Understanding atmospheric pollution
> Looking at problems caused by air pollution

A Air pollution

Burning fossil fuels, oil, coal and gas, creates air pollution (**A**) that is harmful to people and the environment but many countries become developed by using more and more of these fuels. Most developed countries have imposed clean air laws to reduce levels of air pollution. Many developing countries are not able to do this yet. The rapid growth of industry in the developing countries of India and China has caused lots of air pollution and people in cities crowded with factories and cars suffer breathing and health problems. Sixteen of the world's 20 most polluted cities are in China (**B**).

Even in the USA almost half the people live in areas with health-threatening levels of air pollution. The most polluted air is over Los Angeles, partly because of the number of vehicles in use and partly because of the specific location of the city.

IS ATMOSPHERIC POLLUTION ALL THE SAME?

Pollution comes in a variety of different forms.

1. Particles of microscopic soot or dust, produced by power station emissions, diesel exhausts and wood burning, damage your lungs.

2. Smog is mainly made up of ozone and is produced by the reaction of hydrocarbons and nitrogen oxides (produced by burning fossil fuels) in the presence of sunlight. Health problems include eye irritations, breathing difficulties, asthma attacks and bronchitis. Many cities have a haze of smog.

3. Acid rain is made when chemicals of sulphur dioxide and nitrogen oxides combine with moisture in the atmosphere and fall as rain (**C**). Trees are damaged (**D**), the soil contaminated and lakes and rivers become sterile. Efforts are made to control it in Europe and North America where it was first seen but developing countries are now affected, including one-third of the land area of China. One problem is that the acid rain can be carried anywhere, across borders and oceans.

4. Carbon dioxide and other gases are changing the atmosphere and contributing to global warming. Carbon dioxide emissions per capita in metric tons per person between 1990 and 2002 have increased from 1.1 to 2.1 in developing regions and remained the same at 12.6 in developed regions.

B People in polluted area, Nanjing, China

Polluting gases

Acid rain

C How acid rain forms

FACT FILE

WHY IS AIR POLLUTION A PROBLEM?

- We receive 58 per cent of our energy from the air we breathe. How long can you survive without breathing?
- Breathing toxic chemicals damages your health.
- Visibility is reduced by smog and haze.
- Forests, rivers, lakes can be badly affected.
- Buildings, monuments are damaged and discoloured.
- Air pollution is contributing to global warming.

D The results of acid rain in North Carolina, USA

Activities

1 As countries develop, more air pollution is produced. List *five* things you know that create air pollution (look at **A**, **B**, **C**).

2 Aircraft put all sorts of gases into the atmosphere. Should we:

a) pay an extra charge on each ticket towards building cleaner planes?

b) put up ticket prices to stop people flying so often?

c) encourage people to travel by car?

Justify your answer.

3 a) What causes acid rain (**C**)?

b) Acid rain can fall hundreds of miles from where it was made. Why is this a problem?

c) Look at photograph **D**. How has acid rain affected the trees and forest?

d) Chemicals that cause acid rain can be 'cleaned' from emissions before they reach the atmosphere, using extra equipment. Why is this difficult for poor developing countries?

4 Look at photograph **B**.

a) How can people protect themselves against air pollution?

b) As China develops more people are driving cars and fewer are cycling. How will this change affect the atmosphere?

5 List *three* ways to persuade people to use cars less (including one that would stop you driving a car when you are older!), e.g. free buses.

6 Draw a poster suitable for Year 7 students (or make a series of posters with your friends) to show them the causes and dangers of air pollution.

Global warming and development

> **Finding out about the greenhouse effect**
> **Thinking about global warming and development**

Just two gases make up 99 per cent of the Earth's atmosphere – nitrogen 78 per cent and oxygen 21 per cent. Small quantities of other gases make up the remaining 1 per cent but it is these that trap heat around the Earth so they are called greenhouse gases. These gases occur naturally and without them the Earth would be over 30°C colder and uninhabitable! As we increase the quantity of these greenhouse gases, more heat is being retained in the atmosphere, creating global warming (**A**).

Greenhouse gases found naturally in the atmosphere

- Water vapour occurs naturally
- Carbon dioxide (CO_2) produced when people and animals breathe, plants and trees absorb it
- Methane from sheep and cattle digesting food, and from rice paddy fields
- Nitrous oxide (N_2O) from rotting plants
- Ozone occurs naturally

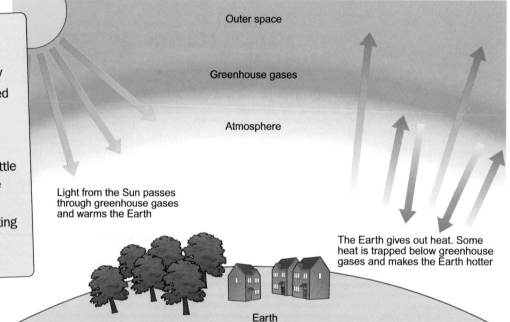

Outer space

Greenhouse gases

Atmosphere

Light from the Sun passes through greenhouse gases and warms the Earth

The Earth gives out heat. Some heat is trapped below greenhouse gases and makes the Earth hotter

Earth

A What causes global warming

Development has increased the quantity of these gases, the balance has changed and the result is an 'enhanced greenhouse effect' or a more efficient greenhouse around the Earth, which is not something we want! Global warming is changing our climates and environments.

HOW DO WE KNOW GLOBAL WARMING IS HAPPENING?

- We can see many changes all around the world, e.g. ice sheets and glaciers melting, plants flowering earlier, sea level rising (**B**).
- We can measure changes – in the atmosphere, in ice cores from polar regions, in pollen samples from the past and in other ways.
- Global temperatures have risen between 0.6°C and 1.0°C in the last 100 years, most in the last 50 years.
- Nineteen of the twenty warmest years recorded in the last 150 years were after 1980.

Although 1°C does not sound very much, world scientists from the Intergovernmental Panel on Climate Change (IPCC) think that if warming increases beyond 2°C our ecosystems and lives will be at risk. The IPPC predicted in 2005 that global temperatures will rise between 1.4°C and 5.8°C from 1990 to 2100.

HOW HAS DEVELOPMENT LED TO GLOBAL WARMING?

- More developed countries use lots of energy from burning fossil fuels, releasing carbon dioxide into the atmosphere. In 2005 the USA produced 24 per cent of the world's carbon emissions.

- Clearing and burning trees adds more carbon dioxide, e.g. Amazon rainforest clearance.

- People in MEDCs eat more meat than those in LEDCs. More sheep and cattle in the world are producing more methane (the biggest single greenhouse gas emitted in New Zealand is methane from sheep).

- Waste landfill sites and increasing numbers of rice paddy fields give off methane.

- Nitrous oxide is increasing as more nitrogen fertiliser is used worldwide.

- Chlorofluorocarbons (CFCs) occur in products which are used more widely now – fridges, plastic foams. CFCs in aerosols have been banned (read the information on the back of an aerosol can).

B Before and after global warming: 1980 and 2003, Andes, Peru

Activities

1 How could planting trees attempt to reduce greenhouse gases?

2 Why does the USA (population 296 million, GDP $40,100) produce more greenhouse gas emissions than China (population 1.3 billion, GDP $5,600)?

3 Draw a table like the one below and complete it.

Gases	Ways in which we are increasing gases
CO_2	
Methane	
N_2O	
CFCs	

4 Suggest ways in which we could produce less of each gas.

5 Global warming information is constantly changing. To find out more about the results of global warming search terms such as 'sea level rise' using Google.

Climate change

> Looking at how we can see climate change happening
> Thinking about the impacts of climate change

A Flooding in Gujarat, India

B Flooding in Carlisle, UK

Greenpeace believes that climate change is the greatest threat facing the planet and that unless we act now millions of people's lives and homes will be at risk. Rising temperatures are producing more droughts, and also floods and storms, which are causing sea levels to rise (**A**).

Climate is the average of daily weather over several years. Information collected about temperature, rainfall, humidity (moisture in the atmosphere) and wind enables weather scientists to plot what has happened to global climates and look for patterns to predict future climates. The evidence collected shows that climate change is taking place more rapidly than scientists had anticipated.

WHAT IS THE EVIDENCE OF CLIMATE CHANGE?

- More frequent and more powerful tropical storms – three of the six most powerful hurricanes recorded in the Atlantic Ocean were in 2005, e.g. Hurricane Katrina, which caused widespread devastation and is the most costly US natural disaster to date.

- Changes in where rain falls and increasing unreliability in when and for how long rain falls.

- Longer dry spells and droughts – in Africa and also lots of other places, e.g. USA, Australia, India, China.

- Changes in direction and strength of winds – high-speed winds are being recorded, as in the tornado that struck part of Birmingham in July 2005.

- Heavier rainfall in a short time – Mumbai, India, July 2005, 944 mm of rain fell in 24 hours, the most ever recorded, floods were widespread and 1,000 people died. Carlisle, UK suffered severe flooding in 2004 and 2005 (**B**).

- Increased heat – possibly 30,000 people died across Europe in 2003 as a result of the hottest summer for 500 years.

WHAT ARE THE IMPACTS FOR PEOPLE AROUND THE WORLD?

All the people in the world are already affected by climate change in some way but those in developed countries have money and resources to deal with the consequences while people in less developed countries do not.

CLIMATE CHANGE IMPACTS

- Ecosystems are changing.

- Flooding is more frequent and dangerous.

- Disease levels may increase, e.g. malaria, spread by mosquitoes.

- Sea level rise threatens coastal areas and many islands. We pay for flood protection schemes in London but Bangladesh, like other less developed countries, has to rely on people building defences such as sandbag barriers.

- Forests, perhaps two-thirds are threatened by changes in rainfall and increased heat.

- Desertification (the spread of deserts) is increasing, often into areas where people want to live and farm (**C**).

- More frequent droughts affect millions of subsistence farmers and may lead to famine.

- Deserts are becoming hotter.

- Unreliable rainfall threatens food production for millions of people especially in LEDCs, but farmers in developed countries may be able to grow different plants.

C Desertification in Burkina Faso. This area was thick with trees fifteen years earlier

Activities

1 a) Why do you think Greenpeace believes that climate change is the greatest threat facing us?

b) Look at the evidence for climate change. Which change do you think will have the greatest effect on the UK? Give your reasons.

2 On a large sheet of paper, draw a spider diagram with nine legs and at the end of each leg describe the impact of climate change. Use illustrations if you can. (See page 155 of *Skills in geography*.)

3 Read the following newspaper extract from July 2005.

a) Name *two* indications of climate change.

> AT LEAST 132 PEOPLE DIED IN FLOODS caused by monsoon rain in the Indian state of Gujarat. More than 400,000 people were left homeless and entire villages submerged. The floods brought other dangers; seven crocodiles up to 1.7 metres long were caught in the city centre and 65 venomous snakes. India has had droughts after poor monsoons recently but this year several months rain fell in a few days.

b) Suggest reasons why it is more difficult for people in India to manage flooding than people in Carlisle.

4 You can find a lot of information about climate change. Try Hotlinks (see page ii) to research extreme climates or extreme weather conditions.

Clean development in the future

> Finding out if development can be less polluting
> Thinking about clean energy and developing countries

Our daily lives in the developed world depend on using energy and most of this energy comes from fossil fuels. More than 80 per cent of the carbon dioxide emissions into the atmosphere each year are energy related. By 2010 consumption of fossil fuels will have doubled since 1990.

One-third of the people in the world have no electricity and burn wood. This damages their health and contributes carbon dioxide to the atmosphere, but people should be able to have a better quality of life and have energy to use. We all need to think about cleaner, better ways to use energy.

Sales of cars in China rose 37 per cent in 2005. Sales in one month alone, September, were a massive 286,400.

4 × 4 SUVs (sports utility vehicles) are called gas-guzzlers as they have very high petrol consumption. They are bad for the environment but loved by people in south-east Asia, the USA and the UK.

FOLLOWING THE LEAD OF THE DEVELOPED WORLD

China is the second biggest energy consumer after the USA but the population is huge so the consumption of energy per person is very low. Most electricity in China comes from burning coal, so if each person used the same amount of energy that we do carbon dioxide emissions would be huge.

CAN WE USE LESS, MORE EFFICIENTLY?

Estimates suggest that 30 per cent of energy used in the UK could be saved by

- turning off lights, televisions etc.,
- using low-energy light bulbs,
- insulating houses,
- driving less, using trains and buses more.

Electric cars cause less environmental damage than petrol cars. Energy-efficient machines use less electricity (look for the sign that tells you the energy rating: 'A' is best).

CAN WE GENERATE ELECTRICITY MORE CLEANLY?

Clean energy, also called green energy, is generated using renewable sources – wind power, wave and tidal power, water power (called hydro electric power, HEP), solar energy, geothermal power (using heat from underground) and biomass (plant material). These all cause less environmental damage than burning fossil fuels.

WHEN THE WIND DOESN'T BLOW

Ten to fifteen years ago scientists in the Netherlands forecast that global warming would cause more storms and wind in north-west Europe. Many wind turbines (**A**) were built in the Netherlands and Germany but the actual number of storms has decreased and the production of clean energy has declined with less wind.

A Wind turbines

SMALL-SCALE SOLUTIONS: CAN IDEAS LIKE THESE SOLVE PROBLEMS?

Decentralised Energy Systems (India) Pvt. Ltd, known as DESI Power, is promoting small rural power plants using renewable energy sources. Suitable, locally grown plants are used in a biomass gasification plant to provide power for the village and rural industry.

The Kenya Ceramic Jiko cooker was designed by local and international agencies. The fuel used is reduced by up 50 per cent, as are toxic gases, so it is healthier to use than old cookers. It is used in half the urban homes in Kenya and has reduced the number of trees used for fuelwood.

Activities

1. If the electricity failed for two days, what problems would you have at home and school (try to think of at least *ten*!).

2. To encourage us to save electricity, would you:
   – have more advertising to tell people to save electricity?
   – charge more for the electricity?
   Give reasons for your answer.

3. Greenpeace has suggested that 4 × 4s should pay a £20 congestion charge to drive in London (because they burn so much fuel and are bad for the environment). Do you agree? Give your reasons.

4. For either a) the rural power generation plant or b) the Kenyan cooker, write a paragraph explaining why this is appropriate for a developing country.

5. Use Google to find out the latest developments in car power other than petrol or diesel, for example using oil from plants. Search 'car fuels'.

● Sustainable development

> **Thinking about sustainable development**
> **Finding out about the Millennium Development Goals**

Sustainability is about living in a way that meets the needs of people today without affecting the ability of future generations to meet their own needs.

- The speed of tropical deforestation is not sustainable because species are being lost forever.

- Increasing the burning of coal as we use it now is not sustainable because carbon dioxide in the atmosphere will increase global warming.

- Using cars as we do is not sustainable because petrol, from oil, is a finite resource and will run out.

- The user-friendly cooker in Kenya and the rural power plant in India are sustainable because they do not irreversibly damage the planet.

Sustainability is an important part of a set of eight goals agreed by world leaders at the United Nations in 2000 called the Millennium Development Goals. The goals work towards improving lives in the developing world.

Millennium Development Goals
The aim of the goals is to ensure that:

- People do not live with extreme poverty and have food and safe water to drink

  - All girls and boys have primary education

- Men and women are equal

  - Few children die before the age of five

- Mothers get healthcare and stay well

  - HIV/AIDS, malaria and other diseases are reduced

- Our use of the environment is sustainable

  - Countries work together for development

MONKEYBIZ AND SUCCESSFUL DEVELOPMENT

Mamotunusi Sekgamane in South Africa can see her life improving as a result of some aid, trade, government help, education and work. In 2000 a non-profit making company, Monkeybiz, was set up, providing beads free to women in the townships then selling their bead work internationally. The project now supports 450 township artists.

A Example of bead work

Mamotunusi is a bead worker. She has five children: the youngest daughter is in high school, one son is disabled from a shooting, another son has TB, one daughter helps her at home and the other son and her husband work away in the city. She likes her house made from wood and corrugated iron, with four bedrooms, a kitchen and now electricity and clean water and she has made it full of colour inside. She is happy that she can get some healthcare for her sons and her own diabetes and a small grant to help her disabled son. Before Monkeybiz she could not work and leave her son at home alone but now she enjoys the beadworking, she earns money to look after her family and lives a more comfortable life. She says 'many people now live in houses made of brick, with electricity and running water so life is much easier now for us in many ways and happier'.

Source: 'Me and My Home: Mamotunusi Sekgamane' by Alice Black. Copyright The Independent, 6 July 2005

Activities

1   Read the Millenium Development Goals, sort them into the order you think most important to achieve and write a summary of them. You could use your own illustration, sketch/drawing or photograph.

2   Use the information above and the article about Mamotunusi Sekgamane to answer the following questions.

- How did local 'aid' help the project start?

- Why do you think selling bead work internationally is important for this project?

- What is Mamotunusi's home like?

- What is her work?

- This project uses local skills. Why is that important for the people?

- Why is it important for her to be able to work at home?

- Why does her family need healthcare and what do they receive?

- What education have her children had?

- How does she describe her life now?

Could she see the development goals being achieved in her life?

Development

Look at the photographs and graphs on this page and the photograph at the beginning of the chapter (page 83).

1 Describe, briefly, the photographs in source **A**.

2 In 2005, 3 per cent of UK energy was from renewable sources. The target for 2020 is 20 per cent. What is the connection between this statement and sources **A** and **B**?

3 Describe the general change in average global temperature (**B**) between 1880 and 1980. Now describe the changes between 1980 and 2004.

4 What is the connection between sources **A**, **B** and **C**?

5 What is the connection between sources **A**, **B** and **D**?

6 There were 815 million hungry people in the developing world in 2002, this was 9 million less than in 1990. Look at source **E**. In which parts of the world has food supply increased? Insufficient food is not just the result of unreliable rain and drought but they are important factors in 34 million more people in sub-Saharan Africa being short of food now compared with 1990. What is the connection between **A**, **B** and **E**?

7 Look at the photograph at the beginning of this chapter (page 83). Find *five* examples that indicate people here have a reasonable quality of life.

8 Use the answers to your questions, and write an article of 200–300 words, headed 'Who is benefiting from development?'

A Sources of carbon dioxide emissions

C Polar bears

D Walking to find water in the desert

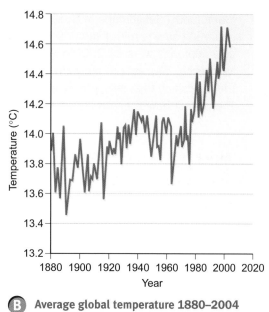

B Average global temperature 1880–2004

Source: Goddard Institute for Space Studies, reproduced at www.earth-policy.org/ Indicators/Temp/TempSmall.gif

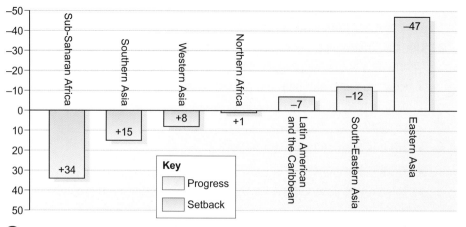

E Change in number of people (in millions) with insufficient food between 1990 and 2002

Source: Millenium Goals Development Report, © 2005 United Nations. Reprinted with permission of the publisher.

» 7 Africa

Africa is a vast continent with a variety of physical and human landscapes. It is a continent of great potential, the challenge it faces is using this potential to improve the lives of the African people.

Learning objectives

What are you going to learn about in this chapter?

> The physical and human characteristics of Africa
> The importance of land as a resource
> The links between education, health and development
> The pressures and challenges of urban growth
> How technology can help countries develop
> The development opportunities of tourism
> The Millennium Development Goals for Africa

A Remote village, Burkina Faso

B Assembly at local school, western Ghana

Africa – the physical environment

> Learning about the physical geography of Africa
> Understanding that Africa has many different natural environments

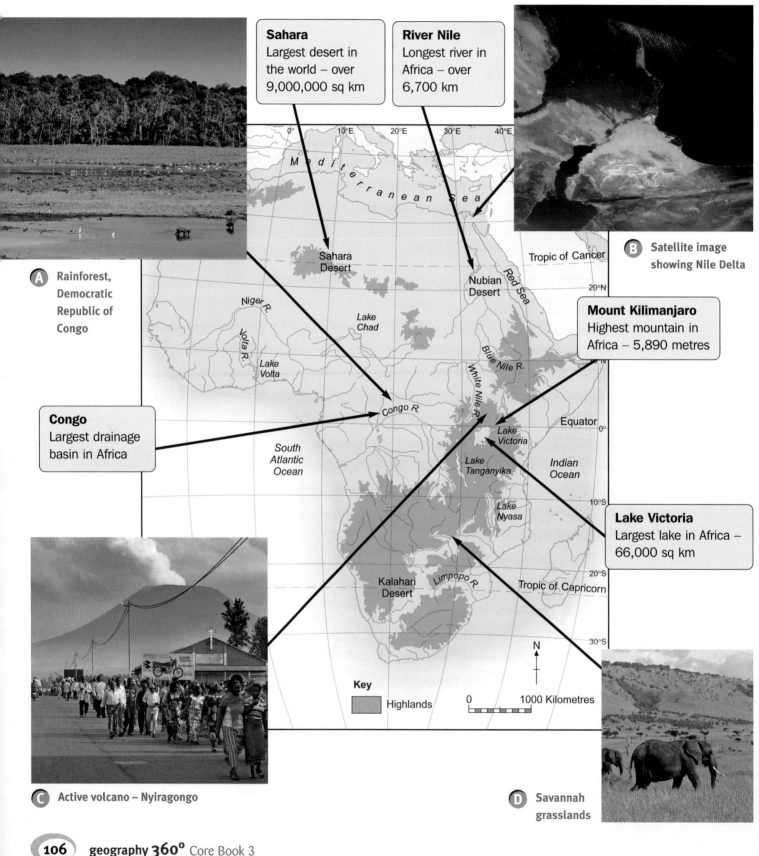

Sahara
Largest desert in the world – over 9,000,000 sq km

River Nile
Longest river in Africa – over 6,700 km

Mount Kilimanjaro
Highest mountain in Africa – 5,890 metres

Congo
Largest drainage basin in Africa

Lake Victoria
Largest lake in Africa – 66,000 sq km

A Rainforest, Democratic Republic of Congo

B Satellite image showing Nile Delta

C Active volcano – Nyiragongo

D Savannah grasslands

Map labels: Mediterranean Sea, Sahara Desert, Nubian Desert, Red Sea, Tropic of Cancer, Niger R., Volta R., Lake Chad, Lake Volta, Blue Nile R., White Nile R., Congo R., Lake Victoria, Equator, Lake Tanganyika, Indian Ocean, South Atlantic Ocean, Lake Nyasa, Kalahari Desert, Limpopo R., Tropic of Capricorn

Key: Highlands

0   1000 Kilometres

N

Africa is a vast **continent** with a lot of different physical environments. It is bounded to the west by the Atlantic Ocean and to the east by the Indian Ocean. The Equator runs through the middle of Africa and the continent stretches to about 36° North and 34° South. This massive range of latitude is one of the reasons why Africa has many different physical environments, including:

- hot deserts with virtually no rain
- savannah grasslands with distinct wet and dry seasons
- rainforests with very high annual rainfall.

Climatic differences also mean that there are differences in plant and animal life throughout the continent.

Africa also has massive rivers, high mountains and active volcanoes – in all it is a continent of great contrasts!

Key words

Continent – a group of countries in the same land mass

Longest rivers	Highest mountains	Largest lakes
1 River Nile	1 Mount Kilimanjaro	1 Lake Victoria
2 River Congo	2 Mount Kenya	2 Lake Tanganyika
3 River Niger	3 Margherita Peak	3 Lake Nyasa
4 River Zambezi	4 Mount Meru	4 Lake Chad

E Facts about Africa

Activities

1   Describe the location of the following features in Africa. Use direction, distance and latitude/longitude if appropriate. The first one is already completed.

Lake Victoria	in East Africa, on the Equator, about 1,000 km from the coast
Nile Delta	
Lake Volta	
Kalahari Desert	

2   Use photographs **A–D** to describe some of the different physical landscapes found in Africa.

3   Use the following data to draw a climate graph for Timbouctou on the edge of the Sahara Desert. (See page 148 of *SKILLS in geography*.)

	January	February	March	April	May	June	July	August	September	October	November	December
Temperature (°C)	20	23	26	30	33	33	32	30	30	29	25	21
Rain (mm)	0	0	0	1	4	19	62	79	33	3	0	0

   a) Calculate the total rainfall for Timbouctou.

   b) What is the temperature range in Timbouctou?

4   **Research task** Draw a map showing the Nile basin. On your map mark and label:

   – Blue Nile   – Lake Nasser   – Mediterranean Sea
   – White Nile   – Nile Delta   – Cairo
   – Aswan Dam   – Red Sea   – the names of the countries that form
                                                                      the Nile basin.

Africa – the human environment

> Learning about the human geography of Africa

> Understanding that Africa has many different human environments

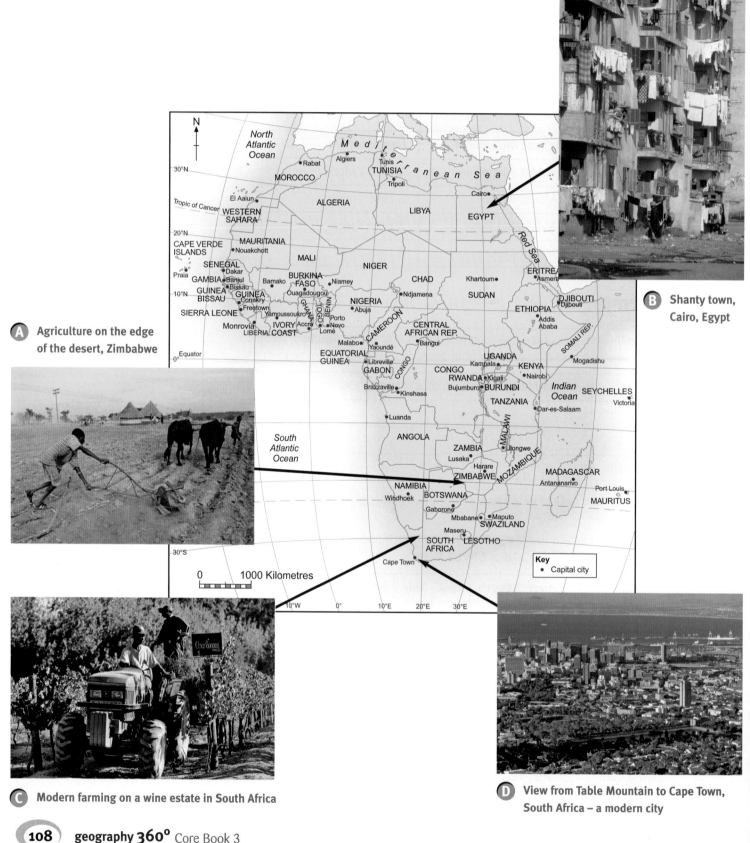

A Agriculture on the edge of the desert, Zimbabwe

B Shanty town, Cairo, Egypt

C Modern farming on a wine estate in South Africa

D View from Table Mountain to Cape Town, South Africa – a modern city

The following statement is from a guide book about Africa.

With over fifty countries, Africa is a vast continent. While some of these countries, like Rwanda, are small, others are huge. Just look at Sudan on a map; it is ten times the size of the United Kingdom!

In days gone by Africa used to be called the 'empty continent' because away from the coast much of the land was uninhabited. Since then the population of Africa has grown but it is still a largely rural continent with many people working on the land. Over the last thirty years some African cities have been growing fast as people move to the urban areas for work. This can be seen in the bustling cities of Cairo, Nairobi, Lagos and Cape Town.

Africa is a continent of great contrasts. Where else can you see age-old **subsistence farming** alongside hi-tech agricultural systems which are providing other parts of the world with fruit, vegetables and flowers?

As the poorest continent in the world, Africa faces great challenges, but it also has great possibilities. With such a variety of landscapes and talented people the desire for progress is strong.

Why not visit Africa and make up your own mind? Take a trip to the ancient pyramids in Egypt, the game reserves of Kenya or the fabulous sub-tropical beaches and vineyards of South Africa; the possibilities are endless.

Largest populations (millions of people)	Largest cities (millions of people)	Wealthiest countries ($ per person per year)	Poorest countries ($ per person per year)
1 Nigeria (137)	Cairo (12)	South Africa (10, 080)	Sierra Leone (500)
2 Egypt (74)	Lagos (11)	Libya (7,620)	Malawi (570)
3 Ethiopia (73)	Kinshasa (6)	Tunisia (6,800)	Tanzania (590)
4 Dem. Rep. Congo (59)	Khartoum (4.5)	Algeria (5,740)	Burundi (620)

E Facts about Africa

Activities

1 Copy out and complete the table on the right by using direction to describe the location of the countries within Africa.

2 Imagine you are planning a trip to Africa. Starting at Cairo you intend to fly to Nairobi and then on to Cape Town. After a few days in Cape Town you then decide to travel to Algiers before returning home.

a) Use the scale bar on the map to measure each leg of the journey in Africa.

b) What is the total distance travelled on the three legs of your journey?

3 Write a paragraph to describe the variety of human characteristics in Africa. Remember to use the photographs!

4 Which of the photographs (**A–D**) do you think is most interesting?

a) Describe the scene in your chosen photograph.

b) Explain why you think it is interesting.

Country	Location
Libya	North
Gabon	West
Mauritania	
Kenya	
Mozambique	
Egypt	

Saving the land – desertification

> Understanding that the land is fragile
> Learning how desertification can be managed

A Preparing rice fields in Mauritania

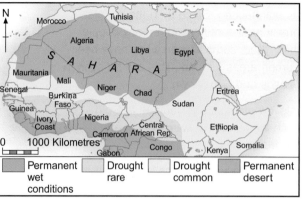

B Where drought occurs in north Africa

In north African countries such as Burkina Faso nearly everybody relies on the land, either for producing food to eat (**A**), or to sell at local markets. Most of the population are subsistence farmers working in communities. They produce crops such as maize, rice, vegetables and fruit, and keep a small number of animals. In some places **cash crops** such as sugar cane and cashew nuts are grown.

WHY IS FARMING DIFFICULT IN NORTH AFRICA?

Conditions for farming are very difficult in areas south of the Sahara desert. This area is known as the Sahel and it has a long dry season and a short wet season. Soils are often poor and can be blown away by the wind during the dry months or washed away when it rains. There is a constant threat of drought (**B**).

WHAT IS DESERTIFICATION?

Desertification is where dry areas become like deserts. In the past 40 years some parts of north Africa have been increasingly affected by desertification (**C**).

WHY DOES DESERTIFICATION HAPPEN?

The causes of desertification are quite complex (**D**) but the results are often simple – less food to eat or sell and, in extreme cases, **famine**.

C Villagers securing sand dunes in an attempt to prevent desertification

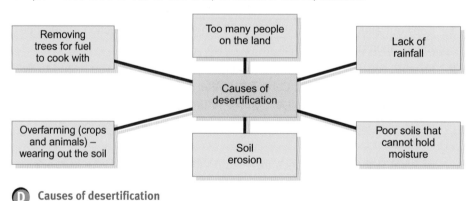

D Causes of desertification

Key words

Cash crops – crops grown to sell, often abroad (exported)
Famine – shortage of food, often resulting in death

REDUCING DESERTIFICATION

The following example shows how a simple project was able to improve farming in one of the poorest parts of Africa.

CASE STUDY   OXFAM IN BURKINA FASO

Why did the people of Yatenga ask Oxfam for help?

Because Oxfam LISTENS

We listened to the people of Yatenga in Burkina Faso. The little rain they had washed the topsoil off their fields. The land they depended on for survival was turning to desert. Oxfam heard about a local method of using lines of stones to dam water and stop topsoil being washed away. By putting lines of stones across a slope, when it rained the topsoil got caught behind the stones and did not get washed away. Also it allowed the rainfall to soak into the ground which meant that crops could get more water. Using this method, local people increased their crop yields by up to 50 per cent.

Because Oxfam TRAINS

Jean-Marie Sawadogo tried Oxfam's method. Seeds planted behind his lines grew well, and crop yields increased dramatically. 'Everyone laughed at me at first,' Jean-Marie remembers. 'But when they saw my crops growing, they started building lines of stones.' To spread the knowledge, Oxfam trainers travel from village to village demonstrating what have become known as 'the magic stones'. Now, more and more people benefit from higher crop yields, and they are improving their land.

Because Oxfam STAYS

Oxfam now teaches people how to grow trees and other plants. As well as providing food, fuel and animal feed, their roots help to hold the soil together which reduces the risk of desertification.

Source: The case study above is adapted by the publisher from the website www.oxfam.org.uk with the permission of Oxfam GB, Oxfam House, John Smith Drive, Cowley, Oxford OX4 2JY, UK. Oxfam GB does not necessarily endorse any text or activities that accompany the materials, nor has it approved the adapted text.

Activities

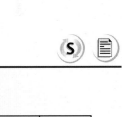

1 a) Use the following information to draw a climate graph for Burkina Faso. (See page 148 of *SKILLS in geography*.)

	J	F	M	A	M	J	J	A	S	O	N	D
Temperature (°C)	22	24	28	32	34	35	32	30	32	31	28	23
Rainfall (mm)	3	3	3	3	5	25	80	80	40	3	2	2

b) Describe the climate of Burkina Faso. Mention the dry and wet seasons and how temperature changes throughout the year.

2 a) How are rainfall, drought and famine linked together?

b) Describe the pattern of drought in north Africa.

3 a) Write a definition of desertification.

b) Explain *three* causes of desertification.

4 a) Use an annotated diagram to explain how the 'magic stones' project in Burkina Faso works.

b) How has the Oxfam farming project in Burkina Faso improved the lives of local people?

5 **Research task** Use the Oxfam website (see Hotlinks, page ii) to describe and explain any *one* farming project in Africa that is improving living conditions.

Increasing food supply in Africa

> Understanding that a lot of people in Africa suffer from a lack of food

> Finding out how local projects in different parts of Africa are increasing food supply

More than one billion people in the world suffer from hunger. Many of them live in Africa. In 2002 it was estimated that over one-third of all children in Africa were **malnourished** and that 40 million African people were **undernourished**.

Poverty and hunger go together because without money it is difficult to get a proper diet, and many people in Africa live in extreme poverty.

HOW IS FOOD INTAKE MEASURED?

Food intake is measured in calories per day. The average person needs about 2,500 calories a day to lead a healthy life, although more is required for people who do hard manual work. On average, people in the United Kingdom have over 3,000 calories a day. In Africa many people have less than 2,000 calories a day (**A**), and consequently suffer from malnutrition.

HELPING PEOPLE IN AFRICA PRODUCE MORE FOOD

Most farmers in Africa are poor, subsistence farmers who cannot afford expensive machinery. In many parts of Africa, low-cost projects using simple technology are helping local people to increase food production. These types of projects are helping thousands of people both to improve their diet and to grow crops to sell. The following examples describe some of the ways that this is happening.

Burkina Faso
The charity Tree Aid is working with women both to increase food output and to grow crops that can be sold. In some areas they have set up mango orchards to produce fresh fruit for local people to eat or to sell in nearby markets. Drying plants are being built so that the fruit can be dried and sold in supermarkets. In other areas, people are being organised to gather shea nuts from wild trees. These are then crushed into oil for cooking or sold as a base for cosmetics.

Zimbabwe
Peanuts are harvested and crushed into peanut butter, which is high in protein. It is used as an infant food and added to stews. The Farmers' Development Trust has set up small processing plants to increase the amount of peanut butter that can be produced.

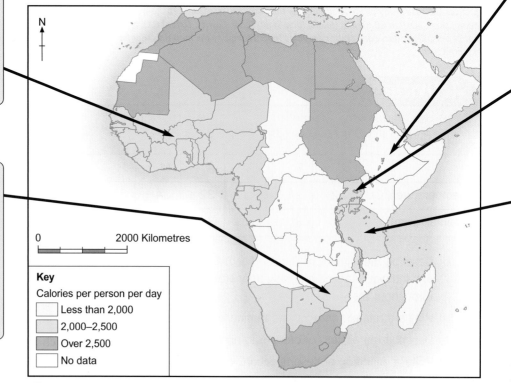

N

0   2000 Kilometres

Key
Calories per person per day
Less than 2,000
2,000–2,500
Over 2,500
No data

A Calorie consumption in Africa

Ethiopia

FARM-Africa is a charity which has a number of projects in Africa. The 'Get your Goat project' is aimed at helping some of the poorest people in rural areas where people often go hungry.

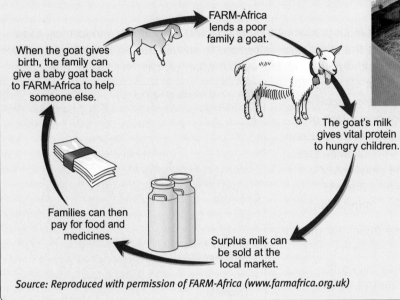

FARM-Africa lends a poor family a goat.

The goat's milk gives vital protein to hungry children.

Surplus milk can be sold at the local market.

Families can then pay for food and medicines.

When the goat gives birth, the family can give a baby goat back to FARM-Africa to help someone else.

Source: Reproduced with permission of FARM-Africa (www.farmafrica.org.uk)

Uganda

Sweet potatoes are an important part of the diet for poor people in Uganda. Recently a virus has been damaging the crop. Scientists working for the Department for International Development (DFID) have produced a new sweet potato plant, which does not catch the virus, has more vitamins and grows faster.

Tanzania

Many people in Tanzania own one or two cows, which produce milk. However, they don't always know how to look after the animals properly. A simple picture guide has been produced to explain how to look after cows more effectively. After using the guide, one village reported that they had cut the cost of keeping their cows by a half but were still getting the same amount of milk.

Activities

1. What is the difference between undernutrition and malnutrition?

2. Why do people who do hard physical work need more calories per day?

3. Name *three* African countries where the average person has:

   a) more than 2,500 calories per day

   b) fewer than 2,000 calories per day.

4. a) Discuss in pairs the link between poverty and hunger.

   b) Write a paragraph to explain the link.

5. a) Explain the negative food cycle (**B**).

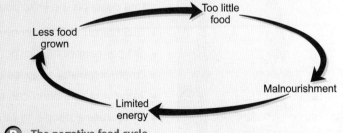

Too little food

Less food grown

Malnourishment

Limited energy

B The negative food cycle

   b) Explain how any one of the projects described on these pages might help to break this cycle.

   c) How might selling excess food improve the lives of poor people?

6. **Research task** Use the FARM-Africa website (see Hotlinks, page ii) to describe *one* project other than the Goat project that is being supported by FARM-Africa.

Water for life

> Understanding the importance of clean water and proper sanitation

> Learning how water supply and sanitation are being improved in parts of Africa

Key words

Contaminated water – dirty or polluted water
Sanitation – system for getting rid of dirty or waste water

Water is essential for producing food and is also used in many industries. Even more important than this: water is needed to sustain human life. We can survive for up to a week without food, but without water we can only live for a few days, even less in hot climates. The problem for many people in Africa is not only a lack of water, but also a lack of clean water (**A**). The following facts sum up the problems of a lack of clean water and proper toilets (**sanitation**).

- Each year about five million people die as a result of poor water supply or **contaminated water**.
- More than one billion people in the world do not have access to safe drinking water.
- More than two billion people in the world do not have proper toilets.

Source: Map reproduced with permission of World Health Organisation (www.who.int/water_sanitation_health/monitoring/jmp04_3.pdf); statistics on sanitation courtesy United Nations Environment Programme UNEP/Grid-Arendal

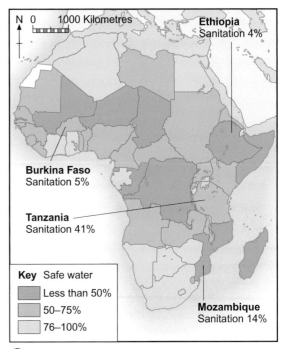

Ethiopia Sanitation 4%

Burkina Faso Sanitation 5%

Tanzania Sanitation 41%

Mozambique Sanitation 14%

Key Safe water
- Less than 50%
- 50–75%
- 76–100%

A Access to clean water and sanitation in Africa

WATER AND DISEASE

In 2004 The World Health Organisation (WHO) stated that 80 per cent of all sickness and disease in poor countries was caused by people having to drink contaminated water and live in places with inadequate sanitation. The following points describe the links between water, sanitation and illnesses like diarrhoea.

- 1.8 million people die every year from diseases linked to diarrhoea.
- 90 per cent of all diarrhoea-linked deaths are children under five.
- Improving water supplies can cut rates of death by up to 25 per cent.
- Improving sanitation can cut rates of death by up to 32 per cent.
- Encouraging people to be more hygienic (hand washing, etc.) can reduce infections by up to 45 per cent.

WOMEN AND WATER

In many parts of Africa women spend hours each day collecting and carrying water (**B**). This not only takes a lot of time but also uses energy.

B Women carrying water, Kenya

If safe water was available near their homes, women could spend more time:

- working to earn money
- improving their education
- growing more food
- improving their homes.

C WaterAid provides a pump in Ethiopia

IMPROVING WATER AND SANITATION IN AFRICA

Ethiopia: the Hitosa water project

This is one of the largest water schemes in Africa. Fresh water is taken from the mountains and piped to over 30 villages in central Ethiopia. Over 50,000 people now have clean running water. Shuma lives in Bekere village, which now has a regular supply of clean water. She said: 'There is less disease and people are much healthier. The girls can now go to school because they don't have to collect water. We have more time for farming and have plenty of water for washing and cooking.'

Mozambique: the family well

'I used to collect water from the swamp which was very dirty,' explains Sara Sanudia, who can now collect safe, clean water for her family from her own well that she built with help from Water Aid. 'Now I have my own well it is much better. It is much closer than before and our health has improved.'

Mozambique: the village toilet

Manuel Oragy is the chief of Muita village. His community have built five new toilets. 'The toilets have made a big difference,' Manuel said. 'Before, we had traditional open toilets. In the rainy season they used to flood and collapse. They smelt very bad too. Now the pits are shallower and they are lined with bricks so they don't collapse. There is no smell and no flies, which is much better.'

D A family well, Mozambique

E An ecological toilet, Mozambique

Source: The information on the three projects is adapted from the WaterAid website (www.wateraid.org.uk)

Activities

1 Draw a spider diagram like the one on the right to describe the different uses of water.

2 Draw a bar graph like the one on the right to show access to sanitation for the countries shown in **A**.

(See pages 155 and 148 of *SKILLS in geography* for help with drawing a spider diagram and bar graph.)

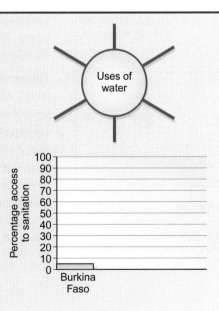

Uses of water

3 **Research task** Using the WaterAid website (see Hotlinks, page ii) produce a one-page presentation about *one* WaterAid project in Africa. Make sure:

– the project is located using a map

– you have described what was done

– you have explained how the project has improved the lives of local people.

The importance of healthcare

> Understanding the links between health and development
> Learning how levels of health are being improved in parts of Africa

According to the World Health Organisation (WHO):

- 20 per cent of the world's population live in extreme poverty
- 30 per cent of the world's children are undernourished
- 50 per cent of the world's population cannot afford necessary drugs.

Poverty is a major cause of ill-health in Africa because the poorest people cannot usually afford enough food, clean water or proper shelter. When poor people are unwell, they cannot always afford to visit a doctor or buy drugs. The link between poverty and ill-health in Africa can easily be seen by looking at the rates of **infant mortality** and **life expectancy** (**A**).

The infant mortality rate in the UK is 5 deaths per 1,000 and the life expectancy is 78 years. Compare this with the countries in Africa shown in **A**.

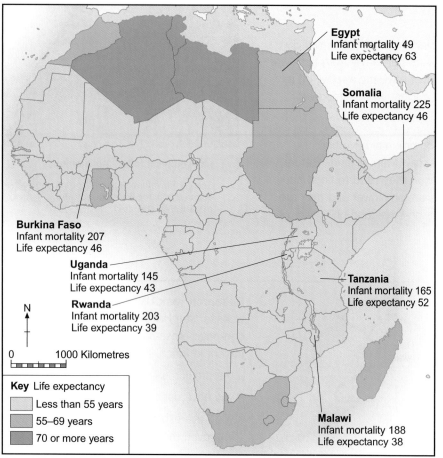

Egypt
Infant mortality 49
Life expectancy 63

Somalia
Infant mortality 225
Life expectancy 46

Burkina Faso
Infant mortality 207
Life expectancy 46

Uganda
Infant mortality 145
Life expectancy 43

Rwanda
Infant mortality 203
Life expectancy 39

Tanzania
Infant mortality 165
Life expectancy 52

Malawi
Infant mortality 188
Life expectancy 38

N

0   1000 Kilometres

Key Life expectancy

	Less than 55 years
	55–69 years
	70 or more years

(A) Infant mortality and life expectancy in some African countries

Source: World Development Indicators 2004. Copyright 2004 by World Bank. Reproduced with permission of World Bank in the format Textbook via Copyright Clearance Center; statistics on life expectancy courtesy United Nations Environment Programme UNEP/Grid-Arendal

WHAT ARE THE MAIN HEALTH PROBLEMS IN AFRICA?

- **Malaria** A disease carried by mosquitoes found in warm, damp climates. It causes extreme fever that weakens the body and can be fatal. Every year millions of people are infected.

- **Malnutrition** Literally 'bad nourishment' or not getting the right food for growth and health. It leaves people with little energy but can also affect eyesight and mental development.

- **Tuberculosis** An infectious disease which affects the lungs. It leaves many people with long-term health problems.

- **HIV/AIDS** AIDS is caused by a virus called HIV (Human Immunodeficiency Virus). It destroys the immune system, which means the body cannot fight infections. The HIV virus is spread through unprotected sex or drug use. It can also be passed from infected mothers to unborn babies.

WORKING TOWARDS A HEALTHIER FUTURE

CASE STUDY　　UGANDA

	Uganda	UK
Adults living with HIV	510,000	34,000
Children living with HIV	110,000	550
Malaria cases per 100,000 people	46	0

B Comparison of health statistics for Uganda and the UK

Uganda's health campaign is making a difference

With the help of the WHO and the United Nations (UN), Uganda is making progress in its fight against disease.

HIV/AIDS The government has started a campaign to give people information about how to avoid infection and where to get treatment for the infection. AIDS information centres have been built and a weekly newspaper has been produced for young people. These measures have helped to halve the number of young Ugandans with HIV in the last ten years, and are helping to push down the rate of other infections such as tuberculosis.

Roll back malaria campaign Malaria is a curable disease and people can protect themselves from infection. Local health centres have been set up to teach people how to protect themselves from the infection or spot it early so they can get treatment before they become ill. Simple use of mosquito nets can make a big difference. The aim is to halve the number of malaria cases by 2010.

Activities (S)

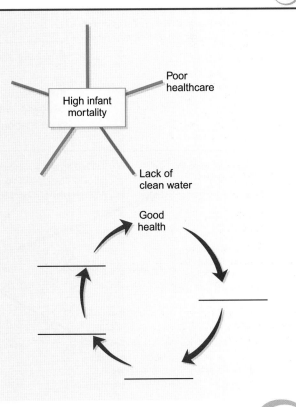

1 a) i) Draw a bar graph to show the infant mortality rates in the countries highlighted in **A** and in the UK. (See page 148 of *SKILLS in geography*.)

　　ii) Why is infant mortality in the UK so low?

　b) Copy out and complete the spider diagram on the right by adding *three* more points that might be suggested by high infant mortality rates.

2 a) Use an atlas or website (see Hotlinks, page ii) to find the number of people per doctor for the UK and for *ten* African countries.

　b) Why are there so few doctors in some African countries?

3 Copy out and complete the circular flow diagram on the right to explain how good health improves life in poor countries.

4 **Research task** Bill Gates of Microsoft has set up the 'Gates Foundation' to improve the health of people in Africa. Look up the 'Gates Foundation' on the internet (see Hotlinks, page ii) and write a paragraph about *one* way it is improving the health of people in Africa.

The importance of education

> Understanding the links between education and development
> Finding out how educational opportunities are being developed in parts of Africa

Nearly one billion people, about a sixth of the world's population, are illiterate and millions of children are denied education every year. In many African countries school attendance rates are low and consequently rates of literacy are also low (**A**). This might be because schools are not available, or are too expensive or simply because children have other tasks to do, such as collecting water. In many African countries there is also a gender difference with far fewer girls attending school.

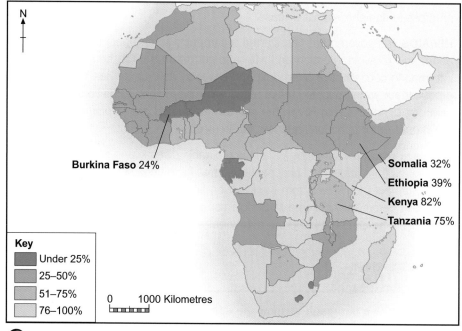

Source: Statistics on literacy reproduced courtesy United Nations Environment Programme UNEP/Grid-Arendal

A Map of Africa showing literacy percentages

Investing in education has been called the 'magic ingredient' in improving people's lives in developing countries. This is because education is not just about qualifications, it can also give young people a number of other advantages (**B**).

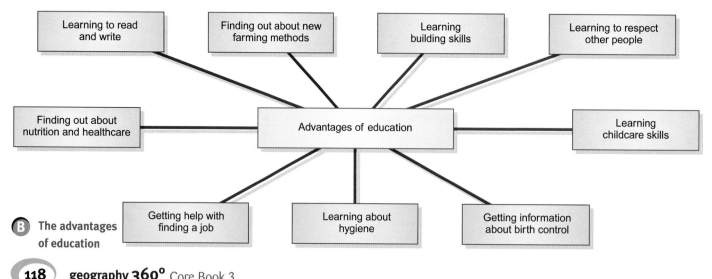

B The advantages of education

'TEENAGE DREAMS'

The following interviews describe the hopes of young people in different parts of Africa.

Milga (17) Somalia
As a girl, Milga has only just been allowed to start school. 'It has been difficult for girls or children from poor family communities to go to school. I am in a class with very young boys. I really want to be a teacher – I shall keep trying.'

Benson (16) Kenya
Benson goes to school and wants to be a doctor or engineer. 'I am lucky because many children cannot afford to go to school. This is a hard place to live, but education may give me a chance, although I may have to leave Kenya to get a job.'

Yannick (13) Burkina Faso
Yannick had to drop out of school when his father died and has to help at home. 'I spend most of my time collecting water from a pond miles away. The water is not clean but it's all there is. A local clean water supply would improve our health and allow me to go back to school.'

IMPROVING EDUCATION IN AFRICA

CASE STUDY   ETHIOPIA'S DIGITAL DREAM

Ethiopia is one of the poorest countries in Africa, and has a shortage of teachers especially in rural areas. 'Schoolnet' is an attempt to bring education to every child in Ethiopia by using Information and Communication Technology (ICT). The Ethiopian government sees it as a great opportunity. A government minister said, 'We want to connect all our villages within three years. ICT may be expensive, but ignorance is even more expensive. In rich countries, computers are often thrown away after two or three years. We could use old computers that still work.' The aim is to use the internet to give up-to-date information and set lessons for the children. The learning language will be English and the aim of the scheme is to improve literacy in some of the poorest parts of Africa.

CASE STUDY   EDUCATION IN TANZANIA

For many Tanzanians, learning to read and write is the way out of poverty. However, over a third of Tanzania's children have never been to school. Often parents cannot afford to buy uniforms, exercise books and pens. Recently the Tanzanian government has announced a plan for 'Free education for all'. At the moment school facilities are poor and there are not enough teachers so parents and friends are helping out. The scheme has meant an extra one million children are in school. With better education, children can learn to read and write and pick up valuable skills. 'I want my children to have an education. It will help them get work, to earn their own living and be able to take care of their family.' (Mwange, a mother of two)

Activities

(S)

1 a) Have a discussion in pairs about how difficult things would be if you could not read or write.

   b) List *six* things that would be difficult if you could not read.

2 Why is it often difficult for children to go to school in parts of Africa?

3 Name two countries in Africa with:

   – under 25 per cent literacy rate

   – 25–50 per cent literacy rate

   – over 50 per cent literacy rate.

4 Explain how any *three* of the advantages of education shown in **B** will improve the life of people in Africa.

5 a) What are the advantages of using computers for education in parts of rural Africa?

   b) What are the difficulties of setting up a computer based education service in a very poor country?

Urban growth in Africa

> Understanding why urban populations in Africa are increasing
> Finding out how life for the urban poor is being improved in some African cities

THE GROWTH OF AFRICAN CITIES

Thirty years ago Africa had few very large cities; perhaps the only one that most people had heard of was Cairo in Egypt. Since that time a small number of African cities have grown rapidly (**urbanisation**), and today there are a number of very large cities across the continent (**A**).

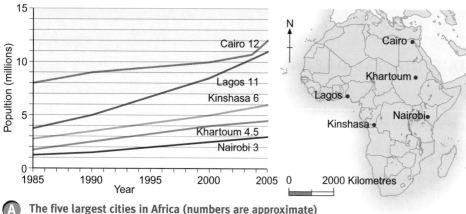

 A The five largest cities in Africa (numbers are approximate)

WHY ARE AFRICAN CITIES GROWING?

The main reason for the growth of population in African cities is **migration**. Every year thousands of people leave rural areas and move to cities because they think the city will give them a better way of life. There are lots of reasons why people move to urban areas and these are often divided into *push* and *pull* factors (**B**).

WHAT ARE THE CONSEQUENCES OF URBAN GROWTH?

Most people who move to urban areas are poor and cannot afford proper housing. They often end up living in slum areas that have very poor conditions (**C**). Life in an urban slum can be difficult – the following paragraph was written by a teenage boy living in Kibera, a slum area with over half a million people in the Kenyan city of Nairobi.

'Life in the slum is hard. Both my parents work long hours but earn little so we cannot afford to move. People keep moving into the area so it gets more and more crowded. Children can go to school here and there is a clinic nearby, but medicine is too expensive for most slum dwellers. We get running water for a couple of hours a day if we are lucky. For many people the street is the toilet so there is often a terrible smell and lots of disease. When it rains the sewage gets washed into the houses and drinking water gets contaminated. There is no waste collection in the slums so piles of rotting rubbish are everywhere – so are the rats! You have to be careful here, crime and violence are never far away.'

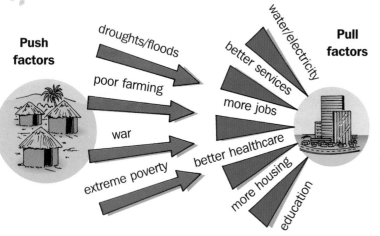

B Reasons why people move from rural areas to cities

C Urban slum in Nairobi, Kenya

IMPROVING THE LIFE OF THE URBAN POOR

The following examples show some of the ways that conditions are being improved in a number of African cities.

Dalifort – Senegal

The 'settlement upgrading project' is improving the basic conditions for slum dwellers. Nearly 100 houses have been given a water supply, electricity and a rubbish collection system. Sanitation systems have been provided for another 7,000 people. The project is encouraging people to improve their houses in other ways.

Addis Ababa – Ethiopia

The government is trying to improve living conditions for the urban poor by:

- giving loans to families so they can build their own homes, either individually or as a small community

- building a small number of basic houses (**D**). These houses have running water and electricity. The rent is low so poor people can afford them.

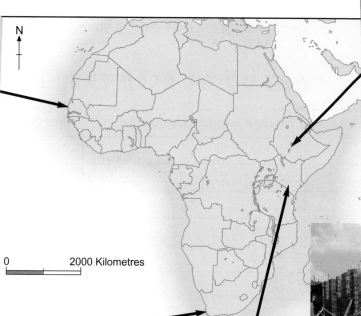

D Basic housing in Addis Ababa, Ethiopia

Cape Town – South Africa

A group of women formed the Guguleto Women's Co-operative (*guguleto* means 'pride'). They formed a savings group and managed to save enough money to buy the materials to build new homes. They then learnt the skills needed to build a small community of new houses.

Kibera (Nairobi) – Kenya

Working with an international charity, the Kenyan government is trying to improve the biggest slum in Africa. It cannot afford to build new homes so has identified the main needs – water and sanitation – as a priority. A local spokesperson said 'clean water and sanitation will reduce disease and stop us having to buy expensive bottled water'.

Activities

1. Draw a table to show the population of the *five* largest cities in Africa in 2005.

2. What is meant by rural–urban migration?

3. a) Working in pairs consider the things that might:
   - make people want to leave an area
   - encourage people to move to a particular area.

   b) (i) Imagine that you are living in a very poor farming area with only basic facilities. Write a brief letter to a friend who lives in a city explaining why you want to leave the countryside and move to the city.

   (ii) Imagine you live in an urban slum. Write to your friend in the countryside explaining that living in the city does have some advantages, but is not always easy.

4. a) Explain how any *two* of the urban improvement schemes might improve life for the urban poor.

   b) Can you think of any disadvantages of your chosen schemes?

Energy and development

> Understanding the link between energy use and development
> Finding out how energy supply is being increased in parts of Africa

WHY DO WE NEED ENERGY?

A Energy use and average income

Country	Average income ($/person/year)	Energy use (kg oil/person/year)
USA	31,000	7,400
UK	24,000	4,100
Brazil	5,200	1,020
Egypt	900	640
Kenya	310	120

The fuelwood crisis in Africa

WOOD IS THE MAJOR SOURCE OF ENERGY for many African families – but it is running out – it is being used faster than it can grow!

This is a real problem for both people and the environment. Some families spend hours every day looking for fuelwood. If they don't find enough, they have to drink un-boiled water and eat raw food – both of which have obvious health risks.

Cutting down trees for fuel can also leave the soil bare. Soil can then be eroded away by heavy rainfall, making farming even more difficult.

B The fuelwood crisis

Looking at the figures in **A** it is easy to see the link between energy use and development. Richer countries use far more energy than poorer countries. They have more industry and transport and also more money to pay for energy.

Imagine how different your life would be if you did not have electricity in your home. In Africa many people have to rely on burning wood as their only source of energy and there are increasing shortages as more and more trees are cut down (**B**).

ENERGY FOR AFRICA

Solar energy in Kenya

'What large parts of Africa need is a reliable and regular source of energy. This would mean more industry, more jobs and better living conditions for ordinary people.' (Government minister)

Very few people in rural Kenya have any sort of power in their homes. Small, portable solar energy collectors were introduced over ten years ago and with the help of local charities their use is increasing. They can be used to power lights, radios and televisions, giving families the chance of getting up-to-date information about health and farming methods. Solar energy is cheap, easy to run and clean, households just need the initial cost of the solar collector (**C**).

C Solar energy collector

50 BILLION DOLLAR PLAN TO USE RIVER CONGO TO PROVIDE ELECTRICITY FOR AFRICA

ONE OF AFRICA'S BIGGEST ELECTRICITY COMPANIES wants to build the world's largest hydro-electricity power station on the River Congo. If it is ever built, it will provide enough electricity for the whole of Africa.

The proposed power station is at the Inga Rapids, near the mouth of the river. It will not need a massive dam, but will involve building large channels where the water can be taken out of the river, passed through turbines to produce electricity and then put back in the river.

The company chairman said, 'This is a really exciting project because it can supply clean energy to some of the poorest parts of Africa. Over 500 million people in Africa do not have electricity and power cuts are common in areas that do have electricity supply.'

Some local people are less enthusiastic about the scheme. A local spokesperson said, '$50 billion dollars could build thousands of smaller energy plants across Africa. Changing the flow of the river will damage the fishing industry and may affect thousands of local farmers who will lose fertile land.'

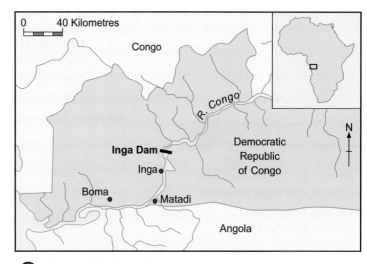

D Course of the River Congo

E River Congo in full flow

Activities

(S) 📄

1 Copy out and complete the spider diagram on the right by adding *five* more uses of electricity in the home.

2 Why is cutting down trees for fuel bad for both people and the environment?

3 Imagine you live in a small village in Kenya and have just had electricity connected to your home. Write a brief report describing the difference it has made to your family.

4 Why are small-scale projects such as solar collectors often more useful than large-scale projects in poor countries?

5 Read the description of the proposed energy scheme on the River Congo.

   a) List the advantages and disadvantages of the scheme.

   b) Do you think it should go ahead? Give reasons for your decision.

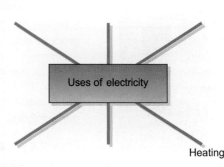

Uses of electricity

Heating

Using appropriate technology for a sustainable future

> Understanding what is meant by sustainable and appropriate development

> Finding out about appropriate development projects taking place in different parts of Africa

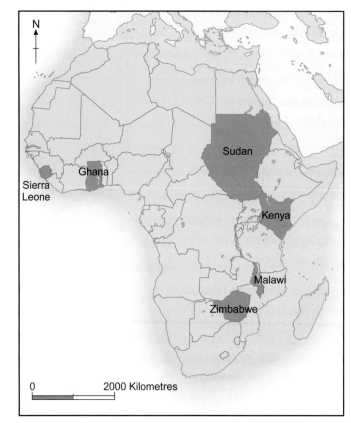

What is development?

The aim of development is to improve the lives of local people. In Africa, many people live in extreme poverty and small changes can make a big difference to the way they live.

The following examples show how small-scale development projects can be **appropriate** and **sustainable**.

A Typical low-income housing, Africa

CASE STUDY

MALAWI – BUILDING STRONG FOUNDATIONS

In Malawi buildings are usually made of 'burnt bricks'. These are made of clay and baked hard in a fire or oven. Trees are cut down to fuel the fires – damaging the environment and leaving less fuelwood for cooking. The process also creates lots of pollution from the smoke of the fires.

Recently a project has been set up to teach local people how to make soil bricks. The bricks are made of local soil and clay, mixed with water. They are pressed into brick shapes in a simple, hand-operated machine for a perfect finish.

Once the bricks have dried out, they are ready to use.

What are the advantages?

The bricks are made of local materials, no heating is needed and they are cheaper and stronger.

KENYA – SUDAN – ZIMBABWE

Practical Action (formerly the Intermediate Technology Development Group) aims to 'find out what people are doing and then help them to do it better'. The following examples of Practical Action projects show how local people are given the chance to start small businesses (**B**).

B Workshop in Kenya

KENYA – 'Practical Action has helped us to produce equipment such as a peanut butter mill to crush peanuts, and fruit juice extractors, which means local farmers can earn more from their crops.'

SUDAN – 'Practical Action has helped to improve the manufacture of metal products by teaching local people new skills and bringing in up-to-date tools. A wider range of products can now be made.'

ZIMBABWE – 'Practical Action has built a number of workshops, which have good tools and machines. Workshop space can be hired and products made to sell in the local area.'

GHANA/SIERRA LEONE/ZIMBABWE

Tools for Self Reliance (TFSR) is a charity that collects old hand tools in the United Kingdom. It then repairs them and ships them out to communities in parts of Africa. The tools are only sent to communities that can really use them. TFSR sets up training courses for young people so that they can become builders, carpenters or tailors. The organisation aims to provide tools and training to help Africans build their own communities.

Activities

1   Explain why the following development projects might not be sustainable:
    – cutting down trees to sell the timber
    – quarrying raw materials to sell abroad.

2   Explain why the Malawi building project is both sustainable and appropriate.

3   a) Working in pairs, discuss how the Practical Action projects might improve the lives of people in Africa.

    b) Write a paragraph to explain your ideas.

4   TFSR collects and repairs hand tools. Can you think of another product that could be collected from richer countries and sent to poorer countries?

    Use a spider diagram to explain how your chosen product would improve the lives of people in developing countries. (See page 155 of *SKILLS in geography*.)

5   **Research task** Use Google to look up *either* Practical Action *or* Tools for Self Reliance. Produce a one-page poster to describe *one* of the projects that the organisation is currently involved in.

Tourism – an opportunity for development

> Understanding that money brought in by tourism can help a country to develop

> Finding out about the different opportunities that Africa can offer to visitors

Key words

National Nature Reserve – area set aside for animals

National Park – area protected because of its environmental quality

A The positive effects of tourism

'Without tourism there would be few jobs in this area. It gives people a better standard of living.'

'Local farmers sell food to hotels and restaurants. Many people make souvenirs to sell to visitors.'

'I work as a hotel receptionist and have been trained to use computers. Meeting holidaymakers has also helped me to learn English.'

'Money from tourism has helped to build a new health clinic. Also a British holiday company has given a lot of money to local schools.'

'Tourism creates a lot of building jobs. Hotels, shops and roads have been built. A new airport and a leisure centre were completed a few months ago.'

The Gambia

The Gambia has miles of beaches and a fantastic climate – just what is needed for a relaxing holiday (**B**). Away from the coast it is possible to visit traditional villages and local markets. Why not visit the famous crocodile pool, followed by a river cruise to spot exotic birds among the tropical forests?

B Beach holiday in the Gambia

Kenya

Kenya offers a range of holiday experiences, from relaxing beach holidays on the Indian Ocean coast to wildlife spotting in the protected **National Parks** and **National Nature Reserves** (**C**).

C Big game in Kenyan safari park

Year	Number of visitors
1985	500,000
1990	800,000
1995	840,000
2000	900,000
2005	950,000

Place of origin	Percentage of visitors
Europe	60%
Africa	25%
America	7%
Asia	8%

D Visitors to Kenya

Tourism is one of the fastest growing industries in the world. Many places rely on the money made from tourism to survive. In developing countries tourism can create jobs and also bring in money for the government (**A**). This money can be used to improve facilities for local people, including healthcare, education or services such as water supply.

TOURISM IN AFRICA

Africa can offer visitors a wide range of opportunities, from traditional beach holidays to action and adventure holidays or wildlife safaris. The following examples from a holiday brochure give just a glimpse of what the continent has to offer.

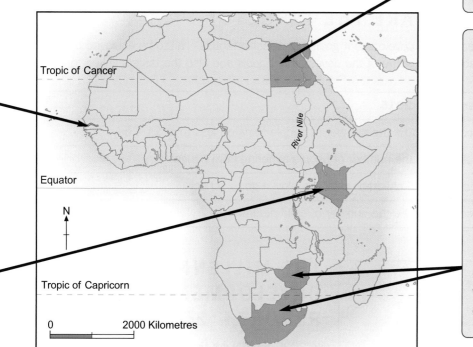

Egypt – Cairo and the River Nile
Cairo and the River Nile have been popular holiday destinations for many years. People visit Cairo to see the pyramids or simply explore the largest city in Africa. Cruising along the River Nile is a relaxing way to see how people live in the riverside towns.

South Africa – Zimbabwe
Southern Africa offers some of the most spectacular scenery in Africa, including Table Mountain in South Africa and the Victoria Falls in Zimbabwe (**E**). South Africa has miles of sandy beaches with many excellent hotels and resorts. Why not visit the wine-growing areas or the wildlife reserves, or even take a short golfing holiday? The Victoria Falls in Zimbabwe is a holiday resort based entirely on the spectacular scenery. Wildlife spotting and whitewater rafting are popular activities.

E Victoria Falls

Activities

1 Make a list of *five* jobs which are linked to tourism:

a) directly (dealing with tourists)

b) indirectly.

2 Explain how the money from tourism can improve living conditions for local people in poorer countries.

3 Draw a table like the one below. For each of the places shown on the map make brief notes describing its attractions.

Place	Attractions
The Gambia	• Miles of beaches • Hot climate • Traditional villages

4 How important is the physical geography in attracting visitors to Africa?

5 a) Draw a line graph to show how the number of visitors to Kenya has increased.

b) Draw a bar graph to show where visitors to Kenya come from.
(See pages 147 and 148 of *SKILLS in geography*.)

c) Imagine that you have spent a few days in a beach resort on the Indian Ocean and then moved to a nature reserve to complete your holiday. Write a postcard home describing your holiday.

6 Tourism brings many advantages to poor countries. Can you think of any possible disadvantages?

The Millennium Development Goals – will Africa achieve them?

> Understanding the aims of the Millennium Development Goals
> Finding out if Africa is likely to achieve the Millennium Development Goals

The Millennium Development Goals were set out by the United Nations (UN) in 2000 (see page 102). The aim of the development goals is to improve the lives of the poorest people in the world.

WHAT ARE THE TARGETS?

Each of the development goals has specific targets – all to be achieved by 2015. Table **A** lists some of the development goals and also the targets linked to them.

A Millennium Development Goals

Development goal	Target – by 2015
A Get rid of poverty	Reduce by 50% the number of people living on less than $1 a day
B Reduce hunger	Cut by 50% the number of people suffering from hunger
C Primary education for all	Make sure all children up to age eleven go to school
D Reduce child deaths	Reduce by 60% the number of child deaths

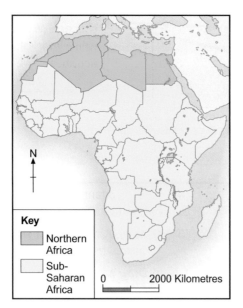

B The two areas of Africa in the Millennium Development Goals Report

Key
- Northern Africa
- Sub-Saharan Africa

0   2000 Kilometres

THE MILLENNIUM DEVELOPMENT GOALS REPORT – 2005

In 2005 the UN set up an investigation to see if the poorest parts of the world were improving. In its report Africa was divided into two areas – northern Africa and sub-Saharan Africa (**B**).

WHAT IS THE UNITED NATIONS?

The UN was formed in 1945. Its aims are to promote peace and look after the needs of all people in the world. Most countries are members of the UN and every country helps to pay for the work it does. In recent years it has done a lot to try to improve the lives of some of the poorest people in the world.

WILL AFRICA REACH THE TARGETS BY 2015?

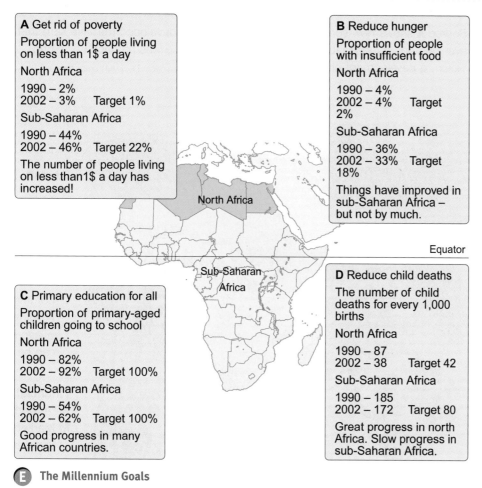

A Get rid of poverty

Proportion of people living on less than 1$ a day

North Africa

1990 – 2%
2002 – 3%   Target 1%

Sub-Saharan Africa

1990 – 44%
2002 – 46%   Target 22%

The number of people living on less than1$ a day has increased!

North Africa

Sub-Saharan Africa

Equator

B Reduce hunger

Proportion of people with insufficient food

North Africa

1990 – 4%
2002 – 4%   Target 2%

Sub-Saharan Africa

1990 – 36%
2002 – 33%   Target 18%

Things have improved in sub-Saharan Africa – but not by much.

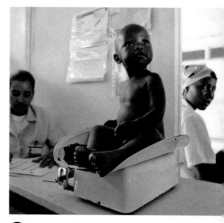

C School children, Kenya

C Primary education for all

Proportion of primary-aged children going to school

North Africa

1990 – 82%
2002 – 92%   Target 100%

Sub-Saharan Africa

1990 – 54%
2002 – 62%   Target 100%

Good progress in many African countries.

D Reduce child deaths

The number of child deaths for every 1,000 births

North Africa

1990 – 87
2002 – 38   Target 42

Sub-Saharan Africa

1990 – 185
2002 – 172   Target 80

Great progress in north Africa. Slow progress in sub-Saharan Africa.

E The Millennium Goals

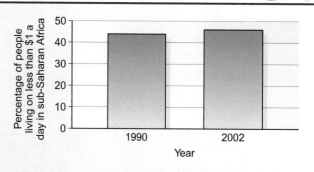

D Health Research Centre, Mozambique

Activities

(S) 📄

1 Produce an information sheet to show how conditions in Africa are changing.

- Use the heading 'Is Africa going to reach the targets for each of the development goals by 2015?'
- Say something about each of the goals mentioned.
- Present some of the figures as graphs.

For example:

A Get rid of poverty

The target for 2015 is 22 per cent in sub-Saharan Africa. At the moment the number of people in sub-Saharan Africa on less than $1 a day is nearly 50% and seems to be increasing not decreasing.

[Bar chart: Percentage of people living on less than $1 a day in sub-Saharan Africa. Y-axis 0 to 50. 1990 approx 44, 2002 approx 46. X-axis label: Year]

2 a) In which area of Africa (see **B**) are living conditions better?

   b) Give *three* reasons for your answer.

3 For which of the goals in **A** are the targets most likely to be achieved?

4 Describe how the lives of African people would be improved if all the targets in **A** were achieved by 2015.

Africa

Source **A** shows how some African villages could be developed to make sure that everyone has a reasonable standard of living. The idea was produced by World Vision, a development charity.

Clean water

A Borehole wells to provide clean water – capable of providing water for approximately 300 people

B Water storage tanks collect runoff from the mountains and provide water for crops and animals

C Earthen dams provide the community with water during dry seasons

Primary healthcare

D Health care centres provide vaccinations and treatment for disease, training for volunteer health workers and parents

E Enclosed toilets help eliminate diseases

Food security

F Livestock farming provides meat, milk and eggs for farmers and their families

G Aquafarms (with water from solar-powered wells) provide fish for healthy diets

H Demonstration nurseries show families the crops, vegetables and fruit-bearing trees they can grow themselves

I Planting trees on bare slopes stops erosion and provides fuel and building materials

J Granaries store grain to ensure a year round supply of food

Education

K Schools enable children to complete primary and secondary school

L Community buildings where village meetings can be held

Financial independence

M Marketplaces give people a place to sell their goods and services

N Bridges built with materials provided by World Vision allow access to other villages and markets

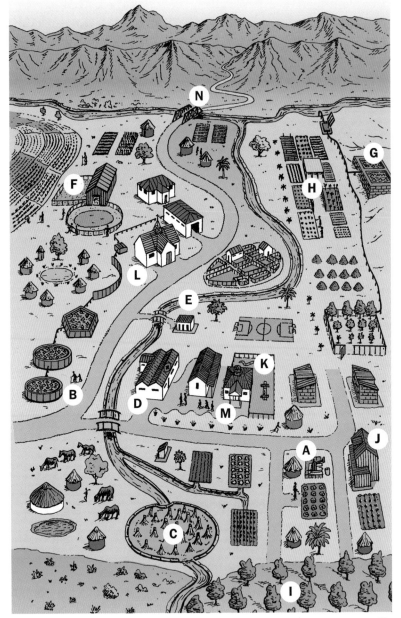

'Vision for a village' has the following aims:
- to reduce disease
- to improve the quality of food
- to make sure everyone has enough clean water
- to improve the ability to read (literacy)
- to create jobs so people can earn money.

For each of the 'Vision for a village' aims write a brief paragraph explaining how the ideas suggested in the diagram will help to meet the aim.

A **'Vision for a village'**

Source: Reproduced with permission of World Vis[...]

Geographical investigations

Learning objectives

What are we going to investigate in this chapter?

> The effects of earthquakes
> Renewable energy in the UK
> Reduce – reuse – recycle
> Street children – homeless children
> Development in South America
> Planning a visit to Africa

A Fridge mountain, Manchester

B Family in Ecuador, South America

C Earthquake in Kobe, Japan

D Remote village, Burkina Faso

People and natural hazards

> Understanding about the effects of an earthquake
> Using a range of presentation skills

Almost every week there are stories in the news about floods, drought, earthquakes and other types of natural hazards (**A**). In geography we study why hazards happen and the effects that they have on people.

A Reports of hazards

The following exercise will give you the chance to show your knowledge and understanding of earthquakes by looking at what might happen if an earthquake struck in Peru, a poor country in South America.

Read the following news report carefully.

News report – Earthquake!

AN EARTHQUAKE HAS OCCURRED IN PERU, a poor country in South America. The epicentre is 100 kilometres west of Lima, the capital city. The shockwaves have brought widespread devastation to Lima. The earthquake measured 7.8 on the Richter scale and early reports suggest great loss of life, damage to buildings and disruption of services.

Thousands of buildings have collapsed and many main roads are blocked. All the city's hospitals are affected. Large areas have no electricity, and water and sewage pipes are broken. Thousands of people are missing, many feared trapped under piles of rubble. Aid in the form of medical supplies, food, tents and clothing is being requested to help this impoverished country.

Your task

You are a journalist who has to produce a front page newspaper report about the earthquake. The editor has asked you to think of a dramatic headline and a story which describes the effect of the earthquake. If there is space, the editor would also like some general background information about earthquakes.

You should work on an A3 sheet of paper and can include maps, diagrams, photographs and any other appropriate style of presentation.

Final note from your editor
Your work must be laid out in the style of a newspaper (look at the examples in source **A** or at some real newspapers).

What should you include?
- A dramatic headline is important.
- Most disaster reports include a map to show the location of the area.
- You will need to write an imaginative report describing what happened and how people were affected. (You could write some quotes from local people.) *Remember –* Peru is a poor country!
- Some explanation about the problems the area might face over the next few weeks would be useful.
- Some background information about why earthquakes happen in particular places would be helpful.

What else could you include?
- Information to show if the area has been affected by earthquakes in the past.
- Some explanation of the Richter scale.
- General background information about Peru or Lima to show what the area is like.
- An article written by an aid agency explaining the types of help the area might need.

Renewable energy in the UK

> **Thinking about the need for renewable energy**

> **Investigating renewables**

Renewable energy is generated using resources that are renewable and not finite like coal, oil and gas (these will run out). They are cleaner because when used they do not add to greenhouse gases in the atmosphere, unlike fossil fuels. The energy is called 'green' or 'clean' and the sources are often just called 'renewables'.

WHY DO WE NEED TO USE RENEWABLES IN THE FUTURE?

- To reduce CO_2 and greenhouse gas emissions into the atmosphere (to slow global warming and climate change).
- To replace/substitute fossil fuels – the UK has used easily accessible gas supplies and is importing gas to fuel power stations.
- The UK target for renewable energy production is 10 per cent by 2010 and 20 per cent by 2020.

A Solar-powered parking meter, Isle of Wight

C Biomass harvesting for wood fuel

WHICH RENEWABLES ARE USED?

- Solar (**A**, **B**)
- Wind (**B**)
- Tidal energy or wave energy
- Geothermal (not to generate electricity but to heat buildings)
- Biomass (**C**) (in several different ways) – yes, burning plants does make CO_2 but growing them takes in CO_2!

B Wind turbine and solar panel, Isles of Scilly

Your task

Your task is to produce an A3 sheet presentation with information referring to renewable energy in the UK.

You are to include at least:

- 1 line graph (e.g. rise in renewable energy production)
- 1 bar graph (e.g. to show change in renewable production 1990 and 2000)
- 1 pie graph (e.g. proportions of types of renewable energy used in the UK)
- 1 map (e.g. map of UK showing location of wind farms)
- details of three types/methods of using renewable sources, e.g. wind, biomass, tidal
- 1 photograph – annotated (well labelled)
- an explanation of why we need to use 'renewables'.

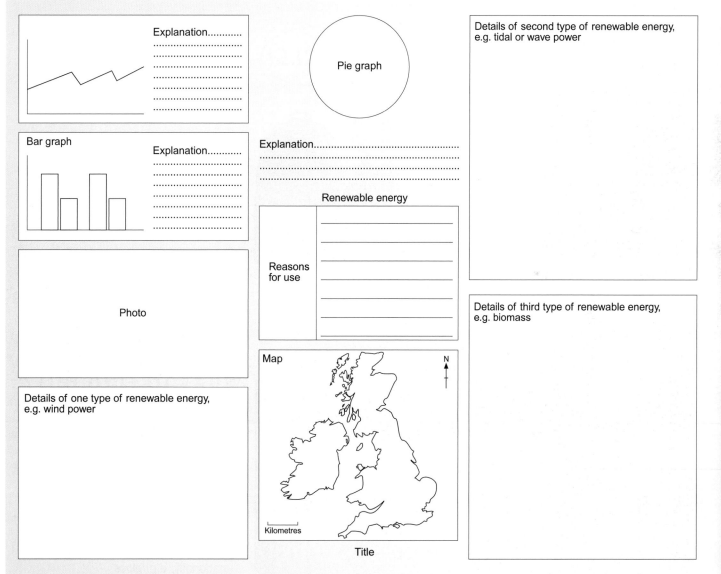

Suggested plan

Search Google and Google Images using key words: for example, 'renewable energy UK', 'tidal power', 'biomass UK'. Three words in a search will be enough but be sure to think about what you want to find out and choose your words carefully.

Reduce – reuse – recycle

> Researching reduce – reuse – recycle
> Selecting and using information for a purpose

Waste disposal is a major problem and expense (UK costs for waste disposal are about £3 billion per year). We put approximately 180 million tons of waste in rubbish bins each year, including:

- about 6 billion disposable nappies
- 972 million plastic bottles
- 468 million batteries.

'Reduce – reuse – recycle' is a slogan used globally. The UK recycled 14.5 per cent of waste in 2002/3 and the target is 25 per cent in 2005/6.

A Reduce your waste!

HOW TO GET PEOPLE TO RECYCLE

- Offer a prize to the person who recycles the most (one authority does this).
- Fine people who do not.
- Can you think of other ways to get people to recycle?

HOW CAN WE GET PEOPLE TO REDUCE WASTE?

- Stop giving out plastic bags in supermarkets.
- Reduce the amount of packaging used on goods purchased in stores.
- Can you think of other ways to get people to reduce waste?

WHAT TO FIND OUT?

- Your local authority waste policy from its website
- Recycling facilities/collections
- What can be recycled and the benefits
- Whats happens to toxic waste
- How long waste takes to degrade, e.g. disposable nappies
- Quantities of things thrown away nationally, e.g. glass
- What recycling takes place in school
- Check out the charity Waste Watch (use Hotlinks, see page ii).

B What can we recycle?

Your task

You are going to produce a leaflet to give to parents, grandparents and staff to persuade people to 'reduce – reuse – recycle'.

- You will explain the benefits of 'reduce – reuse – recycle' and offer them practical ways of doing this.
- The leaflet should be clear, bright and bold with diagrams and illustrations suitable for grandparents, busy parents and staff at school.
- Do include some facts – not too many.
- Make the leaflet eye catching.

You could work out a survey with your group to find out what people do with their rubbish. You might ask:

- How many bags of rubbish do you throw away each week?
- What do you recycle?
- What is collected to recycle?
- Do you throw away food?

Think of at least *three* other questions. The answers will help you plan what to put in your leaflet.

What will the leaflet look like?

A4, either folded in half or in three, with a front cover and your local authority contact details or website on the back.

Front cover with illustration

> Reduce
> Reuse
> Recycle

Front cover with illustration

> Reduce
> Reuse
> Recycle

Include:

- 'reduce – reuse – recycle' logo (see Hotlinks, page ii).
- some facts about UK waste and environmental problems caused by waste disposal
- why we should 'reduce – reuse – recycle'
- what can be recycled
- your local authority waste management details and what is recycled, where and how, for example how they collect food waste
- description of facilities from the authority such as recycling boxes for each household
- a slogan made up by you to encourage people to 'reduce – reuse – recycle'.

PS. If you find lots of information you could make a booklet.

PPS. Can you make enough leaflets in your group to give one to each member of staff?

C Disposable nappies thrown away each year would fill Trafalgar Square ten times over

Street children – homeless children

The United Nations estimates there may be 150 million children living on the streets in the world (**A**, **B**); perhaps half live there twenty-four hours a day. Some work on the streets and go back to their families at night so the exact number of homeless children is difficult to find out.

Key words

Street children – children who live on the streets twenty-four hours a day or who work there all day

A Street beggars in Mumbai, India

Many children are in the rapidly growing urban areas of poorer countries. Millions of AIDS orphans, both parents dead, live in rural areas but are often homeless with no family.

B Children sheltering in a sewer for warmth, Brazil, South America

Poverty is one of the most important reasons children leave home to live rough. Poverty can lead to abuse and certainly to hunger. Why else are children homeless? War and natural disasters are two reasons – can you think of others?

WHAT ARE THE DANGERS ON THE STREETS?

Think what would happen to 9 or 10 year olds alone on the streets in a city faced with violence, drugs, crime.

FINDING INFORMATION

Do a simple search using Google and Google Images with 'street children' and 'homeless children'. You could also try the websites for UN, UNICEF, OXFAM, Comic Relief, Sports Relief, BBC, Consortium for Street Children and Plan International if you have time.

Your task

Your task is to prepare a presentation to give to the people in school who decide how to use money raised for charities. Will this be a school council, a staff group or both? The presentation is to ask for money raised to go to a group supporting homeless children such as Plan International, Comic Relief or Save the Children Fund.

What to include

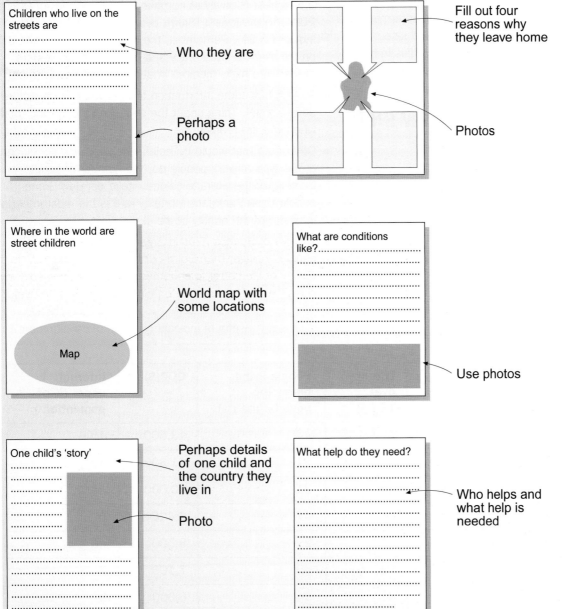

You might decide to work in a group of, for example, four people and take one page each to research. Discuss what information you will need and how and where to find it.

Preparing the presentation

- Decide if you are going to do a presentation on your own or with others.
- Decide whether to do a PowerPoint or spoken presentation.
- How long will it last? Three minutes will seem a long time if you are speaking a presentation.
- Make sure the presentation is appropriate for your listeners – students or staff (adults).

Development in South America

Countries are frequently described using words that are opposites, so they are either MEDCs or LEDCs, rich or poor, developed or developing. A really simple development map divides the world into two – a developed north and less developed south. The continent of South America lies in the south so on the simple map would be called less developed, but is it?

INVESTIGATING DEVELOPMENT IN SOUTH AMERICA

None of the thirteen countries in South America falls into a low human development grouping of the poorest countries in the world and three countries are in a high human development group, so are these three actually MEDCs?

What information might be useful in comparing countries in South America?

- The Human Development Index (HDI) ranks countries from the highest (Norway at number 1, with an index of 0.96 in 2004) to the lowest (Sierra Leone, number 177 with an index of 0.27). Remember, the index is created using data on life expectancy, literacy and GDP and countries are divided into high, medium and low human development.

- Table **A** has some information to help you build a picture of development. Think about the questions around the table when you are reading it.

- Other data that would be helpful include living conditions, or the type of work people do, the number of people farming, or the average calorie intake per day. Some people think the infant mortality rate is the most important indicator of the quality of life in a country.

- The city of Buenos Aires in Argentina with 13 million people is bigger than Bolivia or Uruguay or Paraguay. Is the quality of life better in countries where more people live in cities? Would information on cities and homes be useful?

Is Argentina an MEDC?

Which three countries are in the high human development group?

Is this useful in indicating a level of development?

Country and number in the HDI list	HDI	Life expectancy (years)	Literacy over 15 (%)	People living below the poverty line (%)	GDP($)	Internet users (% of population)
Argentina 34	0.85	76	97	44	13,600	15
Chile 43	0.84	77	96	21	11,300	12
Uruguay 46	0.83	76	98	21	10,000	41
Venezuela 68	0.78	74	93	47	6,400	6
Brazil 72	0.77	72	86	22	8,500	46
Colombia 71	0.77	72	92	55	7,100	7
Peru 85	0.75	70	88	54	6,000	9
Paraguay 89	0.75	75	94	36	4,900	0.3
Ecuador 100	0.73	76	92	45	3,900	1
Bolivia 112	0.69	65	87	64	2,700	0.7

Is Bolivia an LEDC?

How much difference is there between them?

The poverty line is different in each country.

These figures will change frequently.

A Information about development in the ten countries with the biggest populations in South America, 2005

Your task

- Investigate differences in development between countries in South America and think about how development levels can best be described.

- You could make a booklet – or a presentation – to display 'Investigating development in South America'.

- Present the information in annotated maps, graphs and photographs each with a title, key and description of what is shown.

- Use your information to identify and describe differences in development between countries (and within countries where possible).

- Finally, summarise your ideas on development in South America. Will you use the words developed, developing, less developed, rich, poor, MEDC or LEDC?

You could include the following types of information.

- Maps showing countries, cities or population distribution

- A map shading areas to show HDI

Key

	Over 0.80
	0.75–0.80
	0.70–0.74
	Under 0.70

0   2000 Kilometres

- A pie graph showing areas of employment for example

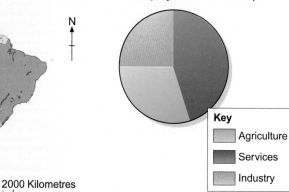

Key

	Agriculture
	Services
	Industry

- Proportional symbols to represent different development levels

Country A   **Country B**

- A table showing different indicators of development for example

	No. of people per doctor	Internet use
Argentina		
Brazil		
Chile		
Ecuador		

- A bar graph showing life expectancy for example

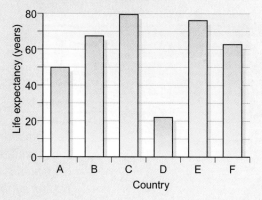

- Annotated photographs

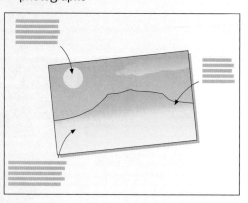

Planning a visit to Africa

> Finding out about different places in Africa
> Using different skills to present information

Africa is a vast continent with a lot to see and many opportunities for activity holidays (**A**). It has a variety of environments, including mountains, rivers, deserts and rainforests. It also has ancient cities, tribal communities and world-famous sites such as the pyramids in Egypt. Some parts of Africa have been popular holiday destinations for many years, including Cairo and the River Nile and the wild animal reserves of southern Africa. More recently the coastal areas of Morocco, Tunisia, The Gambia and Kenya have become popular for relaxing seaside holidays that can also include water sports activities.

THINKING ABOUT A VISIT TO AFRICA

Imagine you are one of a small group of friends who have decided to spend a few weeks visiting different parts of Africa. Source **B** might give you a few ideas to think about.

You need to decide which *five* places you would like to visit, and produce a factsheet describing the attractions of each place. Looking at travel brochures and websites would be a useful starting point. Use Hotlinks (see page ii) to get started.

A Africa

B Thinking about travel in Africa

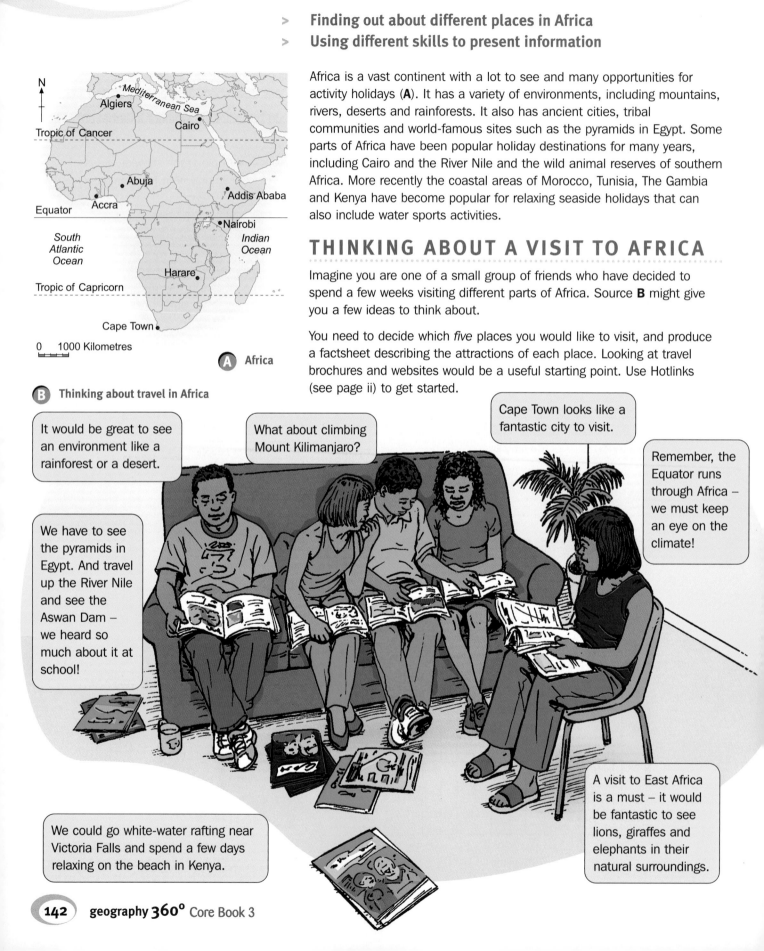

It would be great to see an environment like a rainforest or a desert.

What about climbing Mount Kilimanjaro?

Cape Town looks like a fantastic city to visit.

Remember, the Equator runs through Africa – we must keep an eye on the climate!

We have to see the pyramids in Egypt. And travel up the River Nile and see the Aswan Dam – we heard so much about it at school!

We could go white-water rafting near Victoria Falls and spend a few days relaxing on the beach in Kenya.

A visit to East Africa is a must – it would be fantastic to see lions, giraffes and elephants in their natural surroundings.

Your task

What should your factsheet look like?

Source **C** will give you a few ideas about what could be included and how your factsheet could be presented.

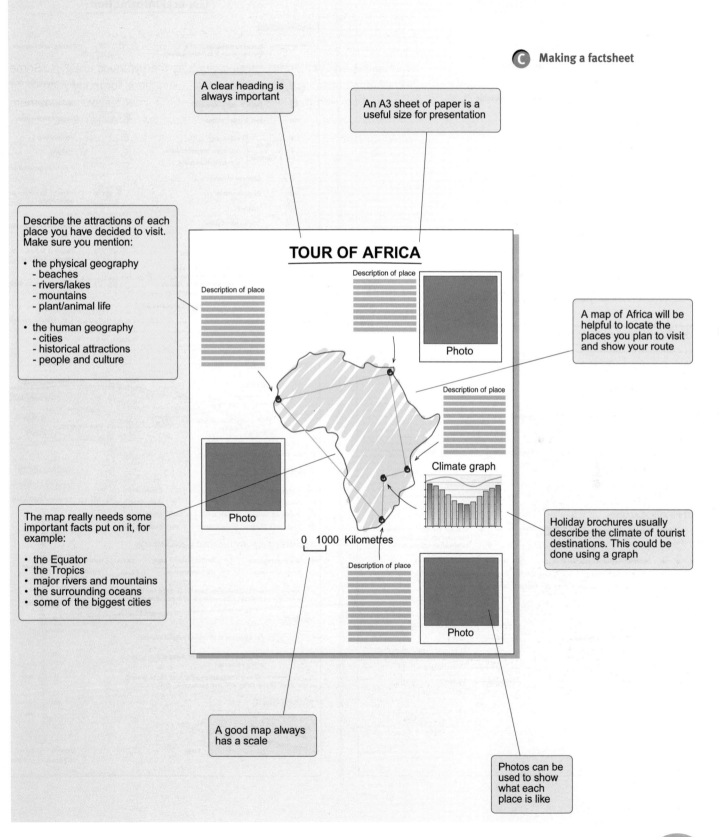

C Making a factsheet

A clear heading is always important

An A3 sheet of paper is a useful size for presentation

Describe the attractions of each place you have decided to visit. Make sure you mention:

- the physical geography
  - beaches
  - rivers/lakes
  - mountains
  - plant/animal life

- the human geography
  - cities
  - historical attractions
  - people and culture

A map of Africa will be helpful to locate the places you plan to visit and show your route

The map really needs some important facts put on it, for example:

- the Equator
- the Tropics
- major rivers and mountains
- the surrounding oceans
- some of the biggest cities

Holiday brochures usually describe the climate of tourist destinations. This could be done using a graph

A good map always has a scale

Photos can be used to show what each place is like

TOUR OF AFRICA

Description of place

Description of place

Photo

Description of place

Climate graph

Photo

0 1000 Kilometres

Description of place

Photo

Ordnance Survey map symbols
(1:50,000)

Communications

ROADS AND PATHS

Not necessarily rights of way

Service area (S) | M 1 | Elevated — Motorway (dual carriageway)

Junction number 1

Motorway under construction

Unfenced | Dual carriageway — Primary Route
A 470

Primary route under construction

Footbridge — Main road
A 493

Main road under construction

B 4518 — Secondary road

A 855 | B 885 — Narrow road with passing places

Bridge — Road generally more than 4m wide

Road generally less than 4m wide

Other road, drive or track

Path
Gradient : steeper than 20% (1 in 5)   14% to 20% (1 in 7 to 1 in 5)

Gates   Road tunnel

Ferry P | Ferry V — Ferry (passenger)   Ferry (vehicle)

PRIMARY ROUTES

These form a network of recommended through routes which complement the motorway system

PUBLIC RIGHTS OF WAY

.... Footpath   ——— Bridleway

— — — — Road used as a public path   -+-+-+-+- Byway open to all traffic

OTHER PUBLIC ACCESS

• • • Other route with public access { not normally shown in urban areas   ◆ ◆ National Trail, European Long Distance Path, Long Distance Route, selected Recreational Routes

● ● ● National/Regional Cycle Network   4 National Cycle Network number

— — Surfaced cycle route   8 Regional Cycle Network number

Danger Area   Firing and Test Ranges in the area. Danger! Observe warning notices.

RAILWAYS

——— Track multiple or single   Station, (a) principal

— — Track under construction   Siding

-+-+- Light rapid transit system, narrow gauge or tramway   —+-○-+— Light rapid transit system station

Bridges, Footbridge   LC Level crossing

Tunnel   Viaduct

General Information

LAND FEATURES

Electricity transmission line (pylons shown at standard spacing)   Cutting, embankment

- - -> - -> Pipe line (arrow indicates direction of flow)   Quarry

ruin   Buildings   Spoil heap, refuse tip or dump

Public building (selected)   Coniferous wood

Bus or coach station   Non-coniferous wood

Place of Worship { with tower / with spire, minaret or dome / without such additions   Mixed wood / Orchard / Park or ornamental ground

o Chimney or tower   Forestry Commission access land

Glass Structure   National Trust-always open

H Heliport

△ Triangulation pillar   National Trust-limited access, observe local signs

Ï Mast

Wind pump/wind generator   National Trust for Scotland

Windmill with or without sails

Graticule intersection at 5' intervals

BOUNDARIES   Administrative boundaries as at May 2004

-+-+-+ National   County, Unitary Authority, Metropolitan District or London Borough

-+-+-+ District   National Park

WATER FEATURES

Marsh or salting   Slopes   Cliff   High water mark

Towpath   Lock   Flat rock   Low water mark

Aqueduct   Canal   Ford   Lighthouse (in use)

Weir   Normal tidal limit   Sand Dunes   Lighthouse (disused)   Beacon

Lake   Footbridge   Bridge   Mud   Shingle

========== Canal (dry)

Contour values in lakes are in metres

ABBREVIATIONS

CH	Clubhouse	CG	Coastguard
MS	Milestone	P	Post office
PC	Public convenience (in rural area)	MP	Milepost
TH	Town Hall, Guildhall or equivalent	PH	Public house

ARCHAEOLOGICAL AND HISTORICAL INFORMATION

+ Site of monument   ✗ Battlefield (with date)   VILLA Roman

•○ Stone monument   ☆ Visible earthwork   Castle Non-Roman

Information provided by English Heritage for England and the Royal Commissions on the Ancient and Historical Monuments for Scotland and Wales

HEIGHTS

—50— Contours are at 10 metres vertical interval

·144 Heights are to the nearest metre above mean sea level

Heights shown close to a triangulation pillar refer to the ground at the base of the pillar and not necessarily to the summit

ROCK FEATURES

Outcrop   Cliff   Scree

CONVERSION

METRES – FEET

1 metre = 3.2808 feet

Metres	Feet
1000	
900	3000
800	2500
700	
600	2000
500	1500
400	
300	1000
200	500
100	
0	0

15.24 metres = 50 feet

SKILLS in geography

1 ATLAS SKILLS

A How to use an atlas 1 – Countries of the world

1 Find the 'Contents' page in the front of your atlas.

2 Look for the heading 'World maps'.

3 Then search for a world 'political map'.

B How to use an atlas 2 – Finding a place

1 Turn to the 'Index' at the back of your atlas.

2 Places are named in alphabetical order.

3 The page for the map you need is given first.

4 Its square is given second.

5 Next its latitude is stated, and then its longitude.

Example:

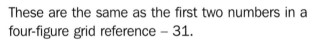

Oxford, UK	5	5E	51° 46´N	1° 15´W
	Page	Square	Latitude	Longitude
			51 degrees	1 degree
			46 minutes	15 minutes
			North	West

The amount of information given, and the order, varies from one atlas to another.

2 OS MAP SKILLS

A How to give a four-figure grid reference

1 Write down the number of the line that forms the left-hand side of the square – the easting – 31.

2 Write down the number of the line that forms the bottom of the square – the northing – 77 (see **A**).

3 Always write the numbers one after each other – do not add commas, hyphens, brackets or a space.

4 Write the number from along the bottom of the map first, then the number up the side – 3177.

B How to give a six-figure grid reference

1 Write down the numbers of the line that forms the left-hand side of the square – the easting.

These are the same as the first two numbers in a four-figure grid reference – 31.

2 Imagine the square is then further divided up into tenths (see **B**). Write down the number of tenths the symbol lies along the line – 319.

3 Write down the number of the line that forms the bottom of the square – the northing. This is the same as the second two numbers in a four-figure grid reference – 77.

4 Imagine the side of the square is divided into tenths. Write down the number of tenths the symbol lies upwards in the square – 774.

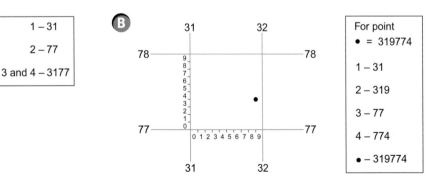

C How to draw a cross-section

When drawing a cross-section from an OS map, you will need to find out the height of the land. See the example below.

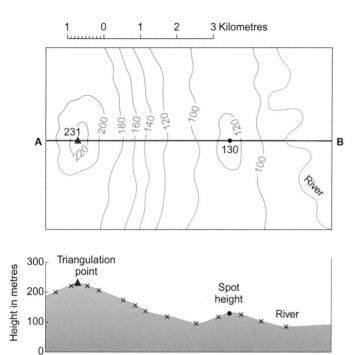

1 Place the straight edge of a piece of paper along the section and mark the start and end point of your section on the paper (AB).

2 Carefully mark on the paper the place where each contour line crosses. Note carefully the heights of the contour lines.

3 Mark on any interesting features, e.g. rivers, roads, spot heights.

4 Now draw a graph of your results. Draw a graph outline (see left). Note the lowest and highest contour height and use this to mark the vertical axis from 0 metres.

NB. Think carefully about the scale up the side – a good guide is 1 cm to 100 m for a 1:25 000 map.

5 Place your paper along the base of the graph and put small crosses on your graph at the correct heights and locations

6 Join the crosses together with a smooth curve – it is best to draw this freehand.

7 Add a title and labels for any key features, e.g. names of hills, rivers and roads.

3 GRAPHS THAT SHOW A TOTAL OF 100 PER CENT

This type of graph allows you to show the parts which make up a total. Think of using one of these four graph types whenever you have to present any data that has a total value of 100 (%). Graphs **A–D** all show the data in the table on the right.

Vehicle type	Number
Buses	20
Cars	70
Lorries	10
Total	100

A ten-minute traffic count near the centre of a UK city

1 Add up the values and make a total.

2 Draw a bar for the total value.

3 The bar can be either vertical or horizontal.

4 Add the scale to the sides of the bar.

5 Plot the different values.

B Pictograph

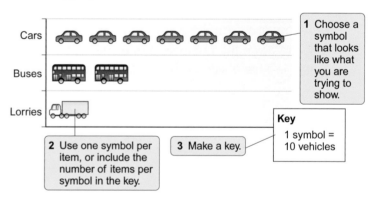

1 Choose a symbol that looks like what you are trying to show.

2 Use one symbol per item, or include the number of items per symbol in the key.

3 Make a key.

Key
1 symbol = 10 vehicles

A Divided bar graph

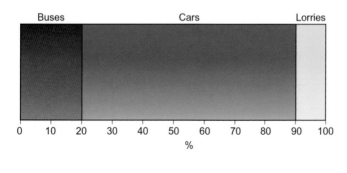

C Pie graph

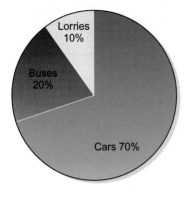

1 If you need to, turn the figures you are using into percentages.

2 Draw a circle.

3 Start at the top (12 o'clock) and draw the segments (from largest to smallest).

4 Make a key or label the segments.

D Block graph

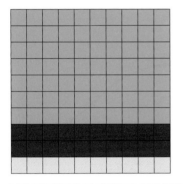

Key

■ Cars

■ Buses

□ Lorries

1 Make a grid of 100 squares. Each square in the block shows 1 per cent.

2 Choose a different shade or colour for each value.

3 Shade or colour in the number of squares for the percentage.

4 Make a key.

4 OTHER GRAPHS

There are many different types of graphs. They are used all the time in geography. Sometimes it does not matter what type of graph you use. At other times, the type of data being shown needs a certain type of graph.

A Line graph

Always use a line graph to show **continuous data**. For example, the only way to show temperature is in a line graph:

J	F	M	A	M	J	J	A	S	O	N	D
4	5	7	10	13	16	18	17	15	11	8	5

Average monthly temperatures in London (°C)

1 Draw the two axes, one vertical and one horizontal.

2 Label what each axis shows.

3 Look at the size of the values to be plotted.

4 Choose the scales and mark them on the axes.

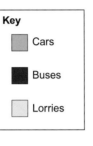

Months of the year

5 Plot the values by a dot or cross.

6 Join up the dots or crosses with a line.

B Vertical bar graph

This graph is useful for showing data that changes every month, or every year. For example, the best way to show rainfall is as follows.

J	F	M	A	M	J	J	A	S	O	N	D
54	40	37	37	46	45	57	59	49	57	64	48

Average monthly rainfall in London (mm)

1 Make a frame with two axes.

2 Label what each axis shows.

3 On the vertical axis make a scale, large enough for the highest number.

4 From the horizontal axis draw bars of equal width.

C Climate graph

How to draw a climate graph

1 Draw graph axes like those in the example here.

2 Allow 12 cm on the horizontal axis for twelve months. Label these J, F, M, etc.

3 Put a scale for rainfall on the lower part of the vertical axis.

4 Above this put a scale for temperature.

5 Plot the monthly rainfall figures as a bar graph.

6 Plot the monthly temperature figures as a line graph. Place each cross or dot in the middle of the column because it is the average temperature for the month.

7 Add a title and label the axes.

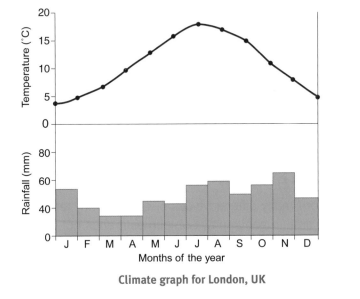

Climate graph for London, UK

How to describe climate graphs

Climate graphs show a lot of data. So, where do you start? The guide below is to help you select the information that is most important. It will make it easier for you to compare the climates of two or more places.

Find the:

1 highest temperature and month

2 lowest temperature and month

3 range of temperature (highest minus lowest)

4 highest precipitation and month

5 lowest precipitation and month

6 precipitation distribution (all year, season with most).

These are the answers for the climate graphs shown.

1 18°C (highest temperature) in July (month)

2 4°C (lowest temperature) in January (month)

3 14°C (range of temperature)

4 64 mm (highest precipitation) in November (month)

5 37mm (lowest precipitation) in March and April (months)

6 All year (there are no dry months)

D Scatter graph

This type of graph is used to show the relationship between two sets of data.

Year	Number of working age (15–64) for every person 65 years and older	Estimates for pension costs as a percentage of GDP
2000	4.3	12.6
2010	3.8	13.2
2020	3.3	15.3
2030	2.8	20.3
2040	2.4	21.4

1 Draw the two axes for the graph.

2 Label the two axes.

3 Choose suitable scales to cover the range of values.

4 Place a cross or dot at the point where the two values meet.

5 Do not join up the dots.

6 If possible, draw a straight line which is the 'best fit' for all the points.

This graph shows a negative relationship (see below). As the number of people of working age **decreases**, pension costs **increase**.

If a relationship exists, it is possible to draw in the line of 'best fit' for all the points. The best fit line is always a straight line. It does not have to go through all the points. It is a summary line which shows the general relationship that exists between the two sets of values.

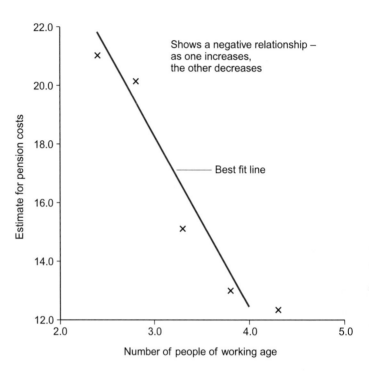

Shows a negative relationship – as one increases, the other decreases

Best fit line

What does the scatter graph show?

The three types of relationship are shown in graphs **A–C** below.

A Positive relationship

As the value of one increases, the value of the other increases as well. Both values increase at the same time.

B Negative relationship

As the value of one increases, the value of the other decreases. One is increasing and the other is decreasing.

C No relationship

It is impossible to see a relationship. The values are scattered all over the place. Drawing a best fit line on the graph is impossible.

Positive relationship

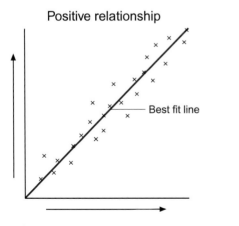

Best fit line

Negative relationship

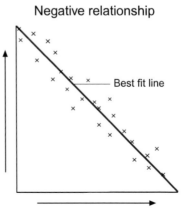

Best fit line

No relationship

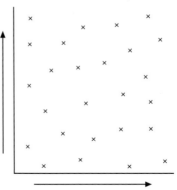

E Population pyramid

1 Show the male population on the left and the female on the right.

2 Draw a horizontal axis with 0 in the middle. The scale can be in either percentages or numbers.

3 Draw a vertical axis from the 0. Divide into age groups, e.g. 0–4, 5–9, 10–14, etc.

4 Draw bars horizontally for each age group and gender.

5 Label the axes and add a title.

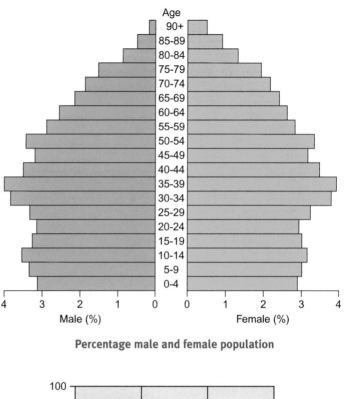

Percentage male and female population

F Block bar graph

Block bar graphs are useful to show two or more values on the same graph. They can be vertical, as here, or horizontal.

1 Draw vertical and horizontal axes.

2 Draw bars for one set of values.

3 Above them draw bars for the second set of values, and so on.

4 Colour in each of the divisions and add a key.

5 Label the axes and add a title.

	Primary	Secondary	Tertiary
India	65	30	5
Japan	8	30	62
UK	4	26	70

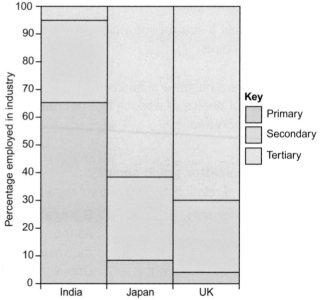

Employment structure in various countries in 2000

G Living graph

1 Draw your graph outline and label the axes. Put the years or time along the bottom and the other value (e.g. *How Laura feels*) up the side.

2 Plot the graph to show how the value changes over time.

3 Add the labels in the correct places on the graph.

4 Add a title to your graph.

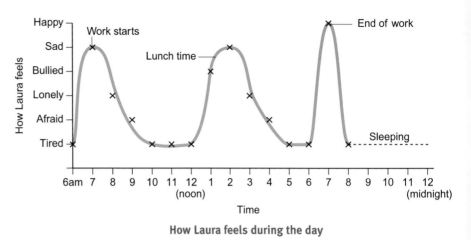

How Laura feels during the day

5 OTHER TYPES OF MAPS

A How to draw a shading (choropleth) map

Shading (choropleth) maps show data for areas. If you have a table of data for named areas of the UK (or for anywhere else), it can be used to make a choropleth map.

1 Look at the highest and lowest values in your table of data, e.g. the highest and lowest wage.

2 Split the values up into four or five groups of equal size.

3 Choose a colour or type of shading for each group.

4 Very important – always choose darker colours for the data groups with the highest amounts (values).

5 Look at your data to see which areas on the map match each value group you set up in step 2. Shade or colour in each area correctly.

6 Remember to add a key!

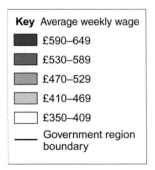

B How to draw a flow map

1 Use a base map showing the places named.

2 Look at the size of the values and the space on your map.

3 Decide on a suitable scale for the width of the lines, e.g. 1 mm for each person or 2 mm for every 5 people, according to the space available.

4 Work out the different line widths.

5 Plot lines of varying width from areas A, B and C to the town.

6 Add a scale and a title to your map.

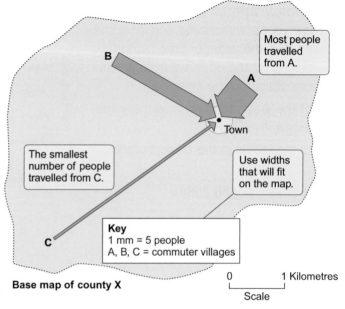

Most people travelled from A.

The smallest number of people travelled from C.

Use widths that will fit on the map.

Key
1 mm = 5 people
A, B, C = commuter villages

Base map of county X

0   1 Kilometres
Scale

Where people travelled from	Number of workers
Place A	40
Place B	20
Place C	5

C How to measure distances

Using the cross-section and the instructions below, you can see that:

The distance between the spot height and the river along line A–B = 1.5 km

The distance between the trig point and the spot height along line A–B = 4 km

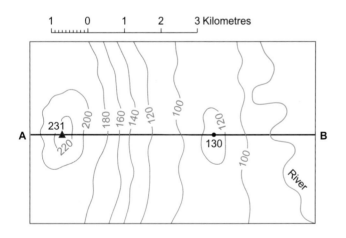

1 Using a piece of paper (or string if it is a winding distance) accurately mark the start and end point of the distance being measured.

2 Transfer the paper or string to the linear scale for the map.

3 Put the left-hand mark on the zero and accurately mark the total number of kilometres on the paper.

4 Measure the bit that is left using the divided section of the scale. This will be in metres.

5 Add the two together to give the final distance measured.

6 Remember to give the units (kilometres or metres) in your answer.

D How to draw a pictorial or mental map

Both types of map use symbols, sketches and diagrams to show locations of geographical features on maps. Below is an example of a pictorial map to show farming. In a pictorial map, the map outline is always accurate and true. In a mental map, the map outline can be accurate, but it may also be distorted.

Pictorial map

This shows what is actually located in an area.

1 Draw or trace accurately an outline map of the area.

2 Make up symbols, sketches, diagrams etc. which look like the features to be shown.

3 Put these in a key

4 Place them on the map where the features are located.

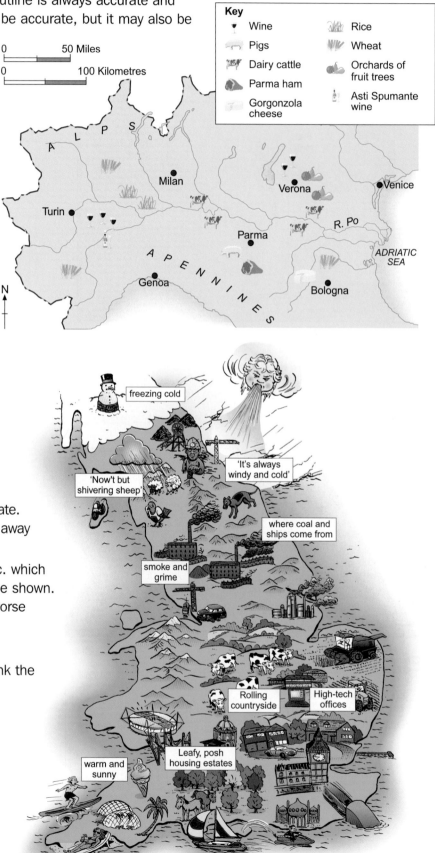

Key		
Wine		Rice
Pigs		Wheat
Dairy cattle		Orchards of fruit trees
Parma ham		Asti Spumante wine
Gorgonzola cheese		

0   50 Miles
0   100 Kilometres

ALPS

Milan
Verona
Venice
Turin
R. Po
Parma
ADRIATIC SEA
APENNINES
Genoa
Bologna
N

Mental map

This shows what people think is located in an area; for geographical features actually located there, it shows what people think they look like.

1 Draw an outline map so that the area can be recognised, even if it is not totally accurate. Remote areas can be made to look further away than they really are.

2 Make up symbols, sketches, diagrams etc. which look like people imagine the features to be shown. Features can be made to look better or worse than they really are.

3 Put these in a key.

4 Place them on the map where people think the features are located.

freezing cold

'It's always windy and cold'

'Now't but shivering sheep'

where coal and ships come from

smoke and grime

Rolling countryside

High-tech offices

Leafy, posh housing estates

warm and sunny

6 SKETCHES

A How to draw a sketch map

The sketch map below was drawn to show differences in relief and drainage.

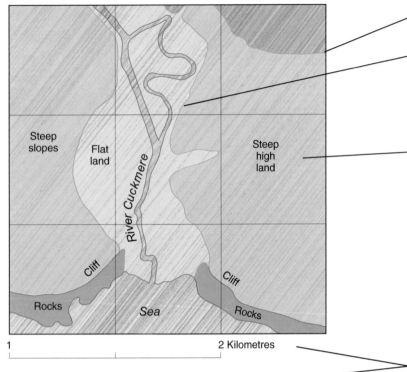

N

Steep slopes

Flat land

River Cuckmere

Steep high land

Cliff

Cliff

Rocks

Sea

Rocks

1   2 Kilometres

Sketch map to show relief and drainage in part of Sussex

1 Draw a frame for your sketch map – think about its size and shape.

2 In pencil, sketch the features you wish to show. Start with some accurate major features such as a coastline or road, or even lightly mark on the gridlines and numbers.

3 Colour in your sketch map. Add a key for the symbols and colours you have used.

Key

▨	River and sea
▨	Woodland
☐	Flat land
▨	Rocks
▨	Steep high land

4 Add a title, north sign and scale.

B How to draw a labelled sketch from a photograph

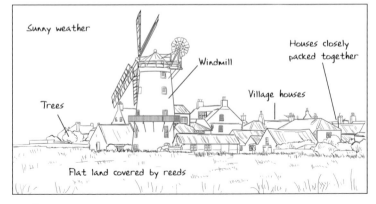

Cley next the Sea, Norfolk

Sunny weather

Windmill

Houses closely packed together

Village houses

Trees

Flat land covered by reeds

1 Make a frame the same size as the photograph.

2 In the frame, draw or trace the main features shown.

3 Label the main physical and human features.

4 Give your sketch a title.

7 TIME COLUMNS

These are used to show what changes happened at key dates in the history of a place. The information contained helps with an understanding of its geography today.

How to make a time column

1 Make a scale of dates down the side of the page.

2 Write in what happened at key dates.

3 Use } for a block of dates and write in what happened.

Date	Information / changes	
1350	Farming village	Farming village
1900	Most houses around the sides of the village green	
1901	Mine opened; terraced houses built around the mine	Mining settlement
1906	Railway station opened	
1957	Mine closed	
1963	Railway station closed	
1971	Motorway built nearby	
1974	Housing estates built	Modern commuter settlement
1994	Becomes a commuter settlement	

8 DIAGRAMS

How to draw a spider diagram

1 Draw a circle (the 'body') in the middle of your page. Write the title in it, e.g. *Factors affecting farming*.

2 Draw lines ('legs') away from the circle.

3 Write a factor at the end of each line, e.g. *Sunshine, Rainfall, Soil*.

4 You could draw a small sketch beside each advantage.

5 You may add as many 'legs' as you need.

6 Each factor, for example rainfall, can have its own set of 'legs' so that you can develop your thinking.

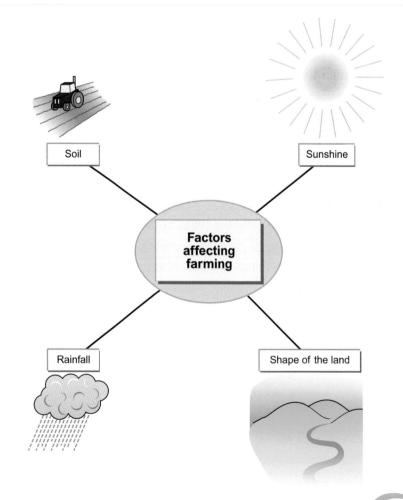

Knowing your levels

Key Stage 3 attainment targets

The targets set out below should help you measure your current level of attainment or realise what needs to be done to achieve your overall Key Stage 3 attainment target.

Attainment descriptions

When looking at your own work or that of your classmates, try to use these bullet points to help you assess the work. Think about what you would need to do to get a piece of work to the next level.

To achieve **Level 3** you also need to:

- Be able to compare the physical and human features of two different places – this means saying what is the same and what is different about places
- Give reasons (explanations) for some of the features found in places
- Give your views about places and say why you have those views
- Talk or write about how people can improve places and maintain their quality for the future
- Answer a range of geographical questions using skills and resources
- Use a range of geographical vocabulary.

To achieve **Level 4** you also need to:

- Study a wider range of places and environments – at different scales and in different parts of the world
- Write about geographical patterns or distributions – be able to say where things are and where they are not
- Describe or write about physical processes e.g. river erosion and human processes e.g. migration

- Begin to understand how these processes can change a place and affect the people who live in an area
- Understand how people can improve or damage environments
- Give reasons for your own views and those of other people about a change to the environment
- Suggest geographical questions and investigate places/environments
- Collect primary and secondary data and present it in different ways and write about what it shows.

To achieve **Level 5** you also need to:

- Explain geographical patterns – why are some things found in some places but not in others?
- Explain the physical and human processes
- Describe how the processes lead to similarities and differences in environments and people's lives
- Recognise some links and relationships that make places and people dependent on others
- Suggest reasons for the ways human activity may change an environment and the different views people have
- Know what sustainable development is and how people try to achieve it
- Explain your own views about environments and places
- Begin to suggest geographical questions and issues
- Investigate places and environments using geographical skills and different ways of presenting information
- Select information, present it, write about it and draw sensible conclusions.

To achieve **Level 6** you also need to:

- Study places at a whole range of scales from local to global

- Describe and explain physical and human processes

- Recognise that processes interact to produce distinctive characteristics of places e.g. the climate in Europe is good for settlement and farming so many people live in this part of the world

- Describe how processes can interact and produce patterns and lead to changes in places e.g. over the last 50 years people have moved out of the centres of cities causing decline in inner cities – the pattern. This has led to a lot of redevelopment in inner cities and people are now moving back – changes.

- Appreciate the links and relationships that make places dependent on each other

- Recognise that there may be conflicting demands on an environment

- Describe and compare different approaches to managing environments

- Appreciate that people, including yourself, have different values and attitudes and that this results in different effects on people and places

- Carry out investigations to answer geographical questions and issues – collect primary and secondary data in different ways, present them using a range of skills, describe and explain what the data show and draw conclusions.

To achieve **Level 7** you also need to:

- Have a knowledge of a wide range of locations and environments at different scales.

- Understand how changes in physical and human processes and interactions may influence the character and distinctiveness of places.

- Understand how different viewpoints influence decisions.

- Explain why places are similar or different and how this affects their characteristics, development and interdependence with other places.

- Understand connections between locations, distributions and patterns of features, why these change and the impact on people and place.

- Understand how interactions at different scales can influence the character and development of places.

- Be able to explain the origins and characteristics of complex issues and evaluate the relevant impact of management structures.

- Understand the factors that contribute to the quality of life and the social, environmental and economic effects of strategies for sustainable development.

- Identify questions and issues, understand the significance of attitudes and values.

- Plan investigations, critically argue points, come to conclusions and evaluate data sources.

- Select and use appropriate skills, accurately and effectively.

Glossary

Appropriate technology technology that can be used effectively by local people to improve their lives

Aquifer rocks holding water underground

Biome a natural large ecosystem named after the main type of vegetation, for example a tropical rainforest

Buttress roots roots stretching above ground from a tall tree to help support it

Cash crops crops grown to sell, often abroad (exported)

Contaminated water dirty or polluted water

Continent a group of countries in the same land mass

Continental Plate plate with land on the surface

Deciduous trees that shed leaves in winter

Desertification arid land being changed to desert due to drought or over-use

Development indicator information to find the level of development of a country, such as infant mortality rates

Drought a lack of rainfall over a long time period

Ecosystem a natural system in which plants, animals, soils and climate are interrelated

Energy source something used to provide energy

Epicentre the point on the earth's surface above where an earthquake starts

Famine shortage of food, often resulting in death

Fossil fuels coal, oil, natural gas, forms of energy created over millions of years, which are non-renewable

Fuelwood any wood that is collected and burnt for fuel, mostly in developing countries

GM crops Genetically Modified crops, crops whose genes have been scientifically changed

Green belt land around cities where building is restricted

Greenfield site land on the edge of a built-up area

High value goods products that sell for high prices such as computers

Illegal logging cutting down trees when it is against the law

Infant mortality number of babies in every thousand born alive that die in the first year

Inner city an industrial and housing area close to the city centre

Interdependence all parts of an ecosystem depend on each other; one change produces other changes

Intermediate technology technology that combines existing skills, affordable tools and new ideas

Irrigation artificially watering landscapes

Landfill site an area where rubbish is collected and dumped then covered over

Lava molten rock on the earth's surface

LEDC less economically developed country

Life expectancy average lifespan at the time of birth

Low value goods things that sell for low prices such as bananas

Magma molten rock inside the earth

Malnutrition lack of the right type of food (vitamins, minerals, etc.)

Mangrove forests forests which grow around coasts where trees are half submerged in saltwater

MEDC more economically developed country

Median age the middle value in a range of ages from youngest to oldest; half the people are younger than this age and half are older

Megacities cities with a population of 10 million people or more

Migration the movement of people

Molten melted

National Nature Reserve area set aside for animals

National Park area protected because of its environmental quality

Natural resource anything we find naturally that we get benefit from, including wood or stone

Non-renewable resource a resource that can be used only once, such as oil

Nutrient cycle the circulation of minerals around an ecosystem

Nutrients minerals such as nitrogen, magnesium, calcium and potassium necessary for plant growth

Oceanic Plate plate with an ocean on the surface

Photosynthesis a biochemical process in which plants use the energy of sunlight to produce food

Plate boundary where the earth's plates meet

Population density a measure of how closely people live together, for example the average number of people per square kilometre

Population distribution how people are spread out over an area

Pull factors things that attract people to places

Push factors things that encourage people to move away from places

Recycled resource something that can be recycled and reused, such as aluminium in drinks cans

Regeneration rebuilding areas so that they attract industry and people

Renewable resource a resource such as water that can be used again and again

Rural an area of countryside with small settlements

Rural–urban fringe the edge of an urban area where it meets the countryside

Rural–urban migration the movement of people from the countryside to urban places

Sanitation system for getting rid of dirty or waste water

Slash and burn the traditional farming used by people in rainforests

Soil erosion removal of soil by wind and rainwater

Street children children who live on the streets twenty-four hours a day or who work there all day

Subduction melting

Subsistence farming a basic type of farming producing food for a family or small community

Suburb an area consisting mainly of housing, outside the city centre

Sustainable development improvement to the lives of people that will last into the future without causing any damage to people or environments

Sustainable resource a resource such as soil and water that can be regularly used so is sustainable but can be over-used or exhausted

Transnational corporations massive global businesses such as Ford, McDonalds or Nike

Tremor shaking caused by an earthquake

Undernutrition not enough food for a healthy life

Urban a built-up area where a lot of people live

Urban sprawl expansion of an urban area into the rural–urban fringe

Urbanisation increase in the percentage of the population living in urban areas

Index